Series in BioEngineering

The Series in Bioengineering serves as an information source for a professional audience in science and technology as well as for advanced students. It covers all applications of the physical sciences and technology to medicine and the life sciences. Its scope ranges from bioengineering, biomedical and clinical engineering to biophysics, biomechanics, biomaterials, and bioinformatics.

More information about this series at http://www.springer.com/series/10358

Amine Nait-ali
Editor

Hidden Biometrics

When Biometric Security Meets Biomedical Engineering

Springer

Editor
Amine Nait-ali
Université Paris-Est, LISSI, UPEC
Vitry sur Seine, France

ISSN 2196-8861 ISSN 2196-887X (electronic)
Series in BioEngineering
ISBN 978-981-13-0955-7 ISBN 978-981-13-0956-4 (eBook)
https://doi.org/10.1007/978-981-13-0956-4

This Springer imprint is published by the registered company Springer Nature Singapore Pte Ltd.
The registered company address is: 152 Beach Road, #21-01/04 Gateway East, Singapore 189721, Singapore

Preface

The term "Biometrics", once split into two halves, highlights the words "Bio" and the word "metrics". These two words should be clearly distinguished since their meaning is deep and precise. The first word stands for life science, whereas the second half stands for measurement. Among the different fields of applications of biometrics, security field has been widely considered and intensively amplified by the media. For this reason, people frequently tend to use the term "biometrics" instead of using "Biometric Security".

Within the context of this book, the focus is on biometric security, and more specifically on hidden biometrics for security. Therefore, before delving into the specifics of biometrics, the concept of classical or conventional biometrics is discussed. Biometric security is defined as a technology which recognizes individuals by their physical or behavioral characteristics. If the modalities considered in this type of biometrics were to be analyzed, we would notice that the "envelope" of the human body and in some cases, its dynamic (behavior) is intensively explored. This means that the role of acquisition devices is mainly to represent or to "Copy" what the human being can perceive, including what the naked eye can see or what the ear can hear. For instance, when the objective is to recognize: face, ears, gait, hand shape, iris, fingerprints, and voice, optical or acoustical sensors are used in most of the cases.

Hidden biometrics for security applications, on the other hand, aims to identify individuals by representing different non-perceived hidden layers of their body and even their hidden behaviors. In order to do so, specific devices and equipment are employed. As discussed in the chapters of this book, some of the devices are mainly used in the biomedical engineering field. Consequently, one can raise the following question: what is the advantage of hidden biometrics compared to the conventional biometrics? The answer is simple, anything which is visible, can be easily accessible, copied, stolen, or imitated. On the other hand, when something is hidden, the access becomes more difficult, hence, it would require more effort, and probably more time to reach the information. So, by considering hidden biometrics, forgeries and attacks become more difficult. This is a big advantage when high-security level is required.

Hidden biometrics can be used to prevent spoofing. For instance, the "brainprint" can be regarded as a signature representing the morphology of our brain which cannot be easily copied. The same observation concerns brain activity, can generally be represented from electroencephalogram. Another advantage of hidden biometrics is the absence of illumination and occlusion issues, commonly observed in the conventional biometrics. In fact, when exploring the inner human body, this issue does not exist.

Although many promising advantages of hidden biometrics are highlighted, we are aware that this technology has "temporarily" some drawbacks. The term "temporarily" is used because we believe that the limitations of today can be overcome in the future. As limitations, one can evoke the acquisition process that can be time consuming, and the cost may be relatively high depending on the modality being used. In addition, user's acceptability factor may be an issue for many reasons, including safety, ethical, and privacy aspects.

Hence, what will be the future of hidden biometrics? What kind of efforts should be deployed? First, it is important to note that technology is being continuously developed, and it would not be surprising that the acquisition systems which are usually used in biomedical engineering will be adapted for security applications. Their size, weight, and cost may be considerably reduced. In fact, since no clinical diagnosis is required, systems can be redesigned and optimized for security purposes. Therefore, if the industry invests in this field in collaboration with experts from the medical field, biometric security can reach a new level. This kind of collaboration is mandatory because this will probably impact users' acceptability. By keeping the users or consumers permanently informed, hidden biometrics could have the same evolution as fingerprint. In the past century, fingerprint identification has been used exclusively for criminals. What happened one century later? People are now using their fingerprint as a key, namely, they can unlock their smartphones with it. Our society is changing and mentalities are changing too, ... So, if you think that the solutions and the approaches presented in this book cannot be exploited nowadays, just keep in mind that tomorrow is another day. You may change your mind, ...

The current book entitled, *Hidden Biometrics: When Biometric Security Meets Biomedical Engineering* is organized as follows:

Chapter "DNA Based Identification": In this first chapter of the book, DNA will be investigated as a deepest Hidden Biometrics modality. After presenting some basic ideas, techniques, and some major applications, a special interest will be addressed to a recent research topics related to the prediction of visible physical traits.

Chapter "A Review on ECG-Based Biometric Authentication Systems": The objectives of this chapter are threefold: First, it presents an overview of the existing ECG benchmarks used for designing ECG-based authentication systems. Second, it presents the literature of authentication systems that used fiducial and non-fiducial features. Third, it presents a methodology that uses both fiducial and non-fiducial features and several data mining classification techniques for individuals' authentication. Moreover, this chapter investigates the pertinent features using a large database of healthy and unhealthy subjects with different heart diseases.

Chapter "EEG Biometrics for Person Verification": The purpose of this chapter is to explore the idea of using EEG signals as a biometric modality to recognize individuals. Considered as a variant of Brain–Computer Interface (BCI), the concept presented in this chapter deals with a Multi-Channel EEG using Emotiv Epoc system. Mainly, a special interest will be addressed to EEG maps analysis for persons recognition. For this purpose, a generic schema is considered, namely, preprocessing, feature extraction, and matching/classification leading to a verification decision.

Chapter "Single Channel Surface EMG Based Biometrics": An emerging biometric method based on Surface EMG (SEMG) signals is considered. For this purpose, this chapter consists of two main parts. The first part reviews the SEMG signals in response to a force of fixed intensity from which frequencial parameters are extracted from the Power Spectral Density (PSD). The second part considers the M-wave signals muscle response following an electrical stimulation. M-wave signals are then characterized by extracting parameters using wavelet networks. The radial basis neural network (RBF) method is then used to classify these parameters. Chapter "Wearable Multi-channel EMG Biometrics: Concepts": In this chapter, a case study using a specific wearable Multi-Channel EMG device will be considered. In particular, eight EMG channels will be used through Myo Armband system. The purpose is to deploy a verification biometric system using EMG signals corresponding to hand gestures. More specifically, the idea behind this concept is the capacity to generate a digital signature for each specific hand gesture.

Chapter "Towards High Density sEMG (HD-sEMG) Acquisition Approach for Biometrics Applications": This is the third chapter of this book dedicated to EMG biometrics modality. The purpose is to highlight a Multi-Channel technique based on a High-Density sEMG (HD-sEMG) acquisition. In fact, HD-sEMG recording systems can be used to overcome the limitations of classical bipolar and monopolar sEMG recording systems. Consequently, in the considered concept, HD-sEMG system generates 64 EMG signals from which an EMG image is constructed and processed. Thereupon, it will be explained how one can deploy this technique in a biometric scheme.

Chapter "Age Estimation Using Sound Stimulation as a Hidden Biometrics Approach": In this chapter, it will be introduced as a new hidden biometrics approach of age estimation requiring the stimulation of the auditory system by an acoustical modulated sine wave signal. After a quick review on different common approaches used in the field of age estimation, and after presenting some generalities on the auditory system, age estimation and age classification protocols will be considered. This chapter describes also the concept of a simple identification/verification, as an application.

Chapter "Multi-, Hyper-Spectral Biometrics Modalities": In this chapter, it will be introduced the different categories of multi-hyper-spectral imaging approaches for biometric modalities. Afterward, indirect approach will be considered, namely, prerequisites concepts on physics and computer graphics, physical theory for the light–skin interaction models and finally, the related applications of multi-hyper-spectral imaging.

Chapter "Imaging for Hidden Biometrics": In this chapter, visible biometrics is highlighted though a classical review on face recognition, including some known approaches from the literature. Therefore, it will be evoked some multispectral approaches such as Near-Infrared (NIR) and Infrared (IR), and in the same context, deep hidden biometrics using X-ray imaging will be considered. Finally, a promising and safe MRI biometrics is described through some recent advances in brain biometrics, which is considered as a robust and safe modality. The purpose of using some hidden biometrics imaging techniques consists basically in preventing potential attacks.

Chapter "Retinal Image Processing in Biometrics": In this chapter, retinal image processing will be addressed as a Hidden Biometric modality. Considered as safe modalities, the retinal vascular network provides a unique pattern for each individual since it does not change throughout the life of the person. In addition, the retina offers a high level of recognition, which makes it suitable for high-security applications thanks to its universality, its invariability over time, and its difficulty to falsify.

Chapter "From Motion to Emotion Prediction: A Hidden Biometrics Approach": In this chapter, it will be discussed the capability of using motion recognition in order to predict the human emotion. Considered as a behavioral hidden biometrics approach, a specific system has been developed for this purpose wherein, several Machine Learning approaches are considered such as SVM, RF, MLP, and KNN for classification and SVR, RFR, MLPR, and KNNR for regression. The study highlights promising results in comparison to the state of the art.

Special thanks to the researchers who contributed to this book and to whom believe that hidden biometrics can be a promising solution of tomorrow.

Vitry sur Seine, France Prof. Amine Nait-ali

Contents

DNA Based Identification

Mohamed Abouelhoda and Amine Nait-ali

Abstract In this first chapter of the book, DNA will be investigated as a deepest Hidden Biometrics modality. After presenting some basic ideas, techniques, and some major applications, a special interest will be addressed to recent research topics related to prediction of visible physical traits.

1 Introduction

As specified in the preface of this book, Hidden Biometrics can be regarded as a particular category of biometrics where the purpose consists of considering Non-Directly-Observable physical or behavioral characteristics of human being. Within this context, DNA is one of the modalities that can be associated to this type of biometrics for the main reason is that, so far, the features extracted from the DNA are not accessible, measured or evaluated by naked eyes as in the case with some other classical modalities (e.g. fingerprint, face, ear, gait, etc.).

The use of DNA as a means for human identification started in the eighties of the previous century, about three decades after the elucidation of DNA structure in 1950 and few years after the introduction of DNA sequencing techniques in 1977. First method for DNA-based fingerprinting was introduced by Sir Alec Jeffreys from the University of Leicester, UK, and concurrently by Dr. Jeffrey Glassberg from Rockefeller University, USA. The major objective of these methods was to confirm the variability in DNA among different individuals and to use this variability to decide if two DNA samples are related or not. The success of Jeffreys and Glassberg

M. Abouelhoda (✉)
Systems and Biomedical Engineering Department, Cairo University, Giza, Egypt
e-mail: mabouelhoda@yahoo.com

Saudi Human Genome Program, King Abdulaziz City for Science and Technology, Riyadh, KSA, Saudi Arabia

King Faisal Specialist Hospital and Research Center, Riyadh, KSA, Saudi Arabia

A. Nait-ali
Université Paris-Est, LISSI, UPEC, 94400 Vitry sur Seine, France

A. Nait-ali (ed.), *Hidden Biometrics*, Series in BioEngineering,
https://doi.org/10.1007/978-981-13-0956-4_1

opened the door for a new field of biometrics based on DNA and biological features. This new field has many applications in forensics, population studies, and criminal investigations.

The last few years have witnessed tremendous advancement in DNA sequencing technology, which lead to more advanced applications beyond measuring the relatedness. These include the prediction of the colors of eye, skin and hair from the DNA, age estimation, prediction of body shape, and even reconstruction of the face.

In this chapter, we will investigate the use of DNA in biometrics. We will explain the basic ideas and techniques as well as the major applications. We will also go through the recent research topics related to prediction of visible physical traits. The coming two sections will briefly explain the structure of DNA and the technologies used to read a DNA sequence. The subsequent sections will address the different techniques in DNA fingerprinting and their applications.

2 The Structure of DNA and Genome

Each cell in the human body includes structures called *chromosomes*. Each chromosome is composed of DNA. The DNA (DeoxyriboNucleic Acid) is a linear chemical compound composed of a chain of chemical units. Each unit is called *nucleotide*. There are only four types of nucleotides in the DNA: Adenine, Cytosine, Guanine, and Thymine; these are abbreviated by the characters A, C, G, and T, respectively. In computer science, one can represent the chromosome/DNA as a string whose alphabet is composed of four characters A, C, G, and T. As shown in Fig. 1, the DNA is not a single strand, it is double stranded in a helix shape. In this double strand structure,

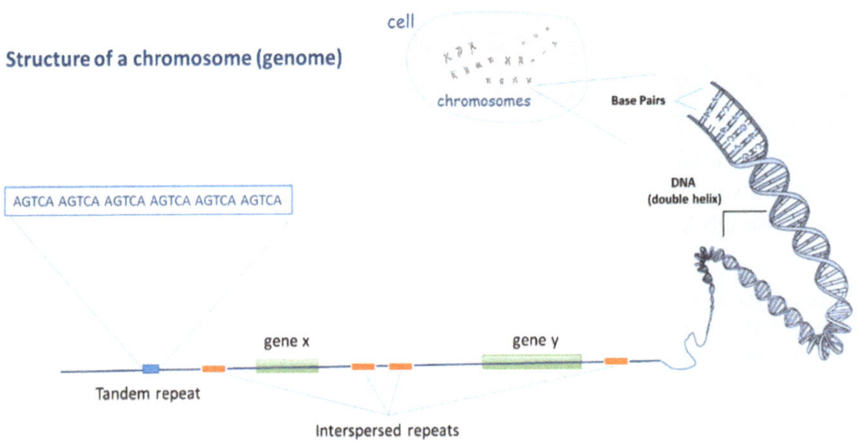

Fig. 1 The cell includes a set of 23 chromosomes. The chromosome is made up of DNA that has double helix structure. The DNA harbors regions that occur one after one another. The regions differ in length, type and order, such as genes and repeats

each nucleotide in one strand pairs (hybridizes) to another nucleotide on the other strand. The pairing follows a strict rule, where A pairs only with T and C pairs only with G. To avoid redundancy, the length of DNA is expressed in terms of *base pairs (bps)*.

The whole DNA (accordingly the whole set of chromosomes) in a cell is referred to as the *genome*. The human genome is composed of 23 pairs of chromosomes. These are referred to as *chromosome 1, chromosome 2,.., chromosome 22, chromosome X, and chromosome Y.* There are two copies of each chromosome (pairs), because one copy is inherited from the mother and the other is inherited from the father. In other words, our genome has 23 chromosomes coming from the mother and 23 chromosomes coming from the father. The total length of the chromosomes in 23 individual chromosomes is 3 billion bps (for the two copies it is 6 billion bps, or totally 12 billion nucleotides considering the double helix structure).

The chromosome does not have a single function. Sub-regions (sub-strings) of the chromosome have different characteristics or functions, and they differ in type, length, and order. The different types of regions can be roughly categorized into (1) Genes, (2) Repeats, (3) Control Sequences and (4) Regions of unknown function. Genes and repeats will be explained in the following sub-sections. Control regions are used to control the synthesis of proteins/RNA in a cell. It is important to note that the order, length, and type of regions differ from one chromosome to another, and they also differ from one organism to another.

2.1 Genes

A gene is a region of the chromosome (the genome) that is used to guide the synthesis of an RNA molecule, through a process called *Transcription*. The produced RNA for some genes may remain as RNA; hence the respective genes are referred to as RNA genes. The other RNAs are further used in another process, called *Translation,* to guide the synthesis of proteins. Hence, the respective genes are referred to as *Coding Genes.* That is, the coding genes are the ones where the end product associated with them are proteins, and the RNA genes are the ones where the end product are just RNA molecules.

RNA (RiboNucleic Acid) is a linear chemical compound similar to the DNA but its length is restricted to the gene length. RNA molecules play different roles in the cell. Some RNAs are used to guide the synthesis of proteins (mRNA), some are used for structural purposes to create some larger molecules (e.g., ribosomal RNA), some play different functional and regulatory roles (like transfer RNA and small interfering RNA).

The protein is a chemical compound which is composed of a chain of amino acids. Once synthesized by the translation process, a protein molecule folds into a unique 3D structure. Proteins play structural roles (like the structure of muscles and hair) and also play functional roles like catalyzing chemical reactions (the enzymes).

The variations in the genes related to human visible traits lead to variations in the appearance of individuals; because these variations lead to differences in the skin, hair, and eye color, and also lead to differences in the body shape and facial traits. The variations in human DNA can be categorized into the following:

- SNP (Single Nucleotide Polymorphism): This is a change of one nucleotide at a certain position in the chromosome. For examples, the nucleotide (base) C changes to A.
- Indels: This includes deletion or insertion of one or few nucleotides.
- Structural variants: This includes the changes where large segments of the chromosome may be deleted, inserted, or moved to another location in the genome.

2.2 Repeated Regions

One of the striking discoveries of the human genome project is that about 50% of the human genome is composed of repeated regions [1, 2]. There is yet no explanation from where these repeats are originated and why they exist. It is just conjectured that they play a role in stabilizing the structure of the chromosome within the cell and protect its terminals from chemical degradation during the life of the organism.

The repeated regions can be classified according to their length, size, and content into the following:

- *Tandem repeats*: These are repeats that occur adjacent to one another. It is a kind of duplication of part of the DNA. The repeated units can be short ones composed of few base pairs or can be long ones composed of hundred or thousand base pairs [3–7]. The shorter ones in the range of few bases are called small tandem repeats (STRs) or microsatellites. The ones that are in the range of 10–100 bps are called variable number tandem repeats (VNTRs) or minisatellites. Micro- and minisatellites are key markers for human identification as will be explained later in this chapter.
- *Interspersed repeats*: These are repeats that are dispersed throughout the chromosome. These are of different types and families. Many of them are thought to be originated from viral sequences which were integrated in the genome. The major families in the human genome include SINE, LINE, LTR, and transposons [1, 2].

3 Reading the Genome

3.1 Sanger Sequencing

DNA sequencing (Genome sequencing) is the process of reading the sequence of nucleotides composing the DNA molecule. The first sequencing method is called

Sanger sequencing. The method is named after the scientist Frederick Sanger who invented it in 1975 [8]. The method starts with cutting the DNA into small fragments. Each fragment is then multiplied into multiple copies using a technique called PCR (Polymerase Chain Reaction). The PCR takes a fragment of DNA, splits the double helix strand into two individual strands, and adds large number of free nucleotides. With the addition of enzymes, the free nucleotides hybridize to the complementary bases in each individual strand, and each individual strand becomes double strand again. The final product after this cycle is to have two identical double stranded DNA fragments. Repeating this cycle, the number of identical DNA fragments doubles each time [8].

The trick in Sanger sequencing is to run four PCR reactions using a set of normal nucleotides and four sets of modified nucleotides. The modified nucleotides terminates the multiplication of a DNA fragment, leading to partial DNA fragment. After the PCR cycles finish, there are DNA fragments of all possible lengths. The fragments can be separated using gel electrophoresis. Because each PCR reaction includes one type of modified nucleotides, it is possible to discriminate or sort the fragments with respect to the nucleotides caused termination.

Sanger sequencing has been later automated using machines that are capable of reading DNA fragments with up to 1000 bps. The high throughput versions of these machines can run up to 96 reactions in parallel. Sanger sequencing is highly accurate and the PCR technique can pick up the fragments in a certain region of the genome. PCR and Sanger method have been used in the sequencing of the first version of the human genome [1, 2]. These methods, with more automation, are still used intensively to sequence the mini- and microsatellite markers used in human identification.

3.2 *Next Generation Sequencing*

Sanger sequencing is slow and expensive to sequence whole genomes or large parts of it. To overcome this limitation, Next Generation Sequencing (NGS) technologies were introduced in the second millennium, and became commercially available for medical applications in the first decade of 2000. After a few years of their appearance, NGS could reduce the cost of sequencing the whole human genome to less than 1000 USD. An NGS machine has high throughput and can read billions of base pairs in one run. The highest throughput model so far (Illumina NovaSeq) can read up to 6 TB of base pairs in one run.

The current major manufacturers of NGS machines are the companies named Illumina and Thermo Fischer. Both companies uses a technology that is referred to as *sequencing by synthesis*. These companies dominate the market, especially for clinical use of NGS.

The basic idea of Illumina sequencing is that it processes millions of DNA fragments in parallel on a special plate called flow cell. When a nucleotide labeled with a fluorescent material hybridizes to a complementary base in a DNA fragment a light is emitted and captured by a CCD camera. Sequence of images of consecutive hybridization events captures different colors corresponding to the sequence of nucleotides in the fragment. After extensive image processing, the real bases composing the DNA fragment are reported [9].

Thermo Fischer uses a similar approach. First, the DNA is multiplied into millions of fragments, which are then processed in parallel. Unlike Illumina that uses fluorescent materials and imaging, Ion technology uses semiconductor technology to sense the change in pH, when a hydrogen atom is released in the hybridization event. Extensive signal processing calls the bases composing the DNA fragments. The basic methods of NGS technologies are firstly introduced in [10, 11].

Although they process millions of DNA fragments in short time, the major problem with the NGS technologies is that only few hundred base pairs from the fragment terminals can be read, which is shorter than what Sanger sequencing can do. For detecting SNPs or Indels this short read length is not a bottleneck, because the large number of reads covering the SNP locus help to differentiate real variations from sequencing errors with high accuracy. However, this short read length limits the ability of NGS to read long micro- and minisatellite regions.

To overcome the short read length of NGS, recent technologies referred to as Single Molecule Sequencing (also called third generation sequencing) have been introduced. This technology can read longer fragments in the range of tens to hundred kilo base pairs. The current limitations of this technology is a higher error rate, but the research is ongoing to improve the accuracy. Current basic manufacturers are Oxford Nanopore and Pacific Biosciences (recently acquired by Illumina).

4 DNA-Based Identification Modalities and Applications

DNA based identification modalities can be broadly categorized into *genotyping* based and *phenotyping* based. Genotyping is about the use of selected DNA markers (certain SNPS, microsatellite or minisatellite regions) which are proven to be highly variable among individuals. Genotyping markers usually do not encode any RNA or protein and do not have any control function. Phenotyping is about using other DNA markers from different visible traits (phenotype) of the human. The variations in these markers are related to changes in human visible traits like pigmentation (hair, eye, and skin color), body shape, age, and facial features.

Genotyping can be used in different applications, based on measuring the level of relatedness of two or more samples. It can be used in population studies to figure out, for example, how different human populations evolve and relate to each other. Such studies can be based on using minisatellite markers [12, 13] or SNP markers from whole genome sequencing [14–19]. Relatedness measures can also be used in forensics to investigate paternity or to decide whether a suspect's DNA matches the DNA found in crime scene.

The early use of genotyping techniques was in forensics and criminal investigations [20, 21]. The first example was the case of confirming the paternity of a child to his migrant mother in an immigration dispute. This case was solved by the research group of Prof. Alec Jeffreys from Leicester [22]. Another success by Jeffreys' group was confirming the identity of a murderer whose DNA was collected at the crime scene and was arrested by the police, after collecting around 5000 samples from local men. This success opened the door for using DNA-based testing as a reliable means in forensics.

Genotyping used in population genetics was also pioneered by the group of Sir Jeffreys. The group published many articles showing the relatedness among different population groups and could confirm the African origin for modern humans [12]. Following similar approaches, many studies have appeared afterwards. To take a few examples, Jobling et al. studied the African population [13], Alonso et al. [23] studied the Basque population and could prove their European origin, Rogers et al. [24] studied the Eurasian population, Brión et al. [25] studied the relation between the European lineages and the North African one, and Yuan et al. [26] studied the Thai, Chinese, and Japanese populations.

An interesting example of the power of genotyping is their use in solving the dispute on the fatherhood of the US President Jefferson to a son of his slave. Using DNA samples from male-line descendants of Jefferson, Foster et al. could prove that Jefferson is the biological father of her last son [27]. This example could demonstrate how powerful DNA analysis is, not only for solving recent cases but also for answering historical and archeological questions.

Although phenotyping based techniques are very novel and still in research phase, they were used to solve few forensic cases. The first version of these techniques was used by the company Parabon (https://snapshot.parabon-nanolabs.com) in 2015 to help the law enforcement authorities of Rockingham County, USA, to solve a cold double-murder case which took place in 2012. The company provided the authorities with a report predicting the skin color, the eye color, and the ancestry of the suspect (European/Latin). This information guided the authorities to focus the research and identify the murderer. For this analysis, the company used the SNP array CytoSNP-850 K chip from Illumina, which included over 850,000 SNPs.

After this primary success, the company further developed the technology to produce what is called Snapshot® DNA analysis, which can also reconstruct the face, in addition to the color of eye, hair and skin. Although the prediction of the face is still approximate, it still helps the authorities narrow down the search and solve some cases. The algorithms used by Parabon are not published and it is not yet known how accurate the prediction is over large dataset.

Interestingly, the prediction of visible traits have also been used in very recent archeological studies to reconstruct the face of the "Cheddar Man" [28], the Jomon Woman in Japan, and the Ava from Caithness [29].

5 Genotyping Based Techniques

5.1 RFLP Analysis

The first technique used to roughly compare two DNA samples to figure out if they are from the same person or not was based on the Restriction fragment length polymorphism (RFLP) technique. In this technique, the DNA is cut at certain locations using a set of restriction enzymes. The individual variations at the cut locations of the enzymes lead to variable length DNA fragments. The DNA fragments can then be placed on a sheet of Gel and separated by electric current. Each person would have a pattern different from one another.

5.2 Repeat Based Analysis: Minisatellites

Sir Jeffreys was the one who discovered that minisatellite repeats show high variability among individuals and can be used to identify persons. Minisatellites were used in different criminal investigations as shortly presented above and as explained in [20].

A minisatellite locus is a region in the chromosome that is composed of multiple repeated sub-regions. Each of these sub-regions is called a unit. These units may not be of identical sequence, and may include some variations. The different unit types are given names or symbols (typed). In Fig. 2, we show the five types in the minisatellite locus MSY1 in chromosome Y. The linear arrangement of the minisatellite units in one locus in one person is referred to as a minisatellite map. In computer science, it can be represented as a string, where the alphabet is the unit types. As shown in Fig. 2, the minisatellite maps differ from one person to another either in the type of units included or the number of repeated units.

Manual comparison between two minisatellite maps is possible to figure out if the two maps in study are related or not. But for larger datasets, automatics methods have been developed to align and compare satellite maps [30]. The WAMI server [31] provides a visual alignment between pairs of maps and constructs a phylogenetic tree showing the relatedness among multiple maps.

The use of minisatellites has a number of challenges for large scale fingerprinting. First, it is sophisticated and difficult to automate. Second, it is more expensive, and time consuming. Since the early 1990, the forensic practice favored the use of microsatellites [32]. The coming sub-section includes explanation of the microsatellites and the related techniques.

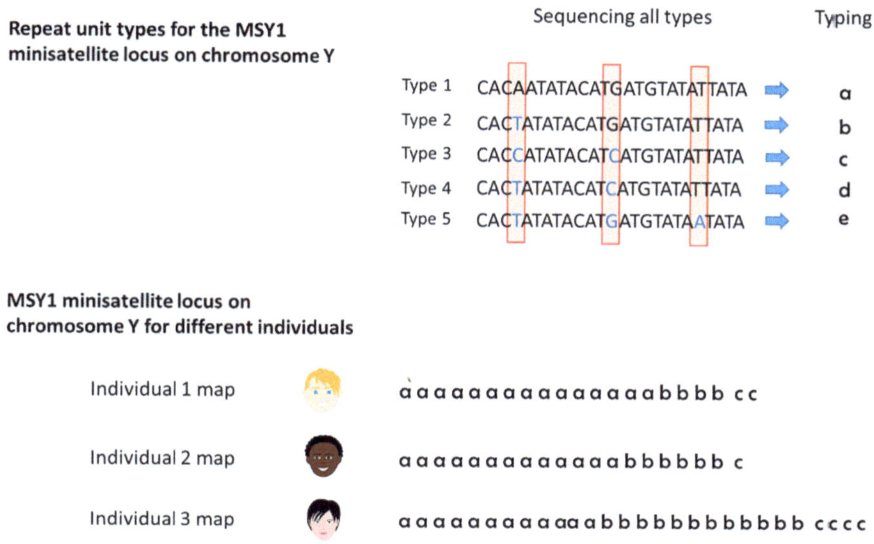

Fig. 2 An example of mini-satellite locus (MSY1) on the Y chromosome. The upper part includes the unit types composed the mini-satellites at that locus. The lower part shows examples of mini-satellite maps from different individuals

5.3 Repeat Based Analysis: Microsatellites

Microsatellites, also referred to as short tandem repeats (STRs), are important markers for human identifications. The number of repeated units in a microsatellite locus vary from one individual to another. Microsatellites have gained much popularity in human identification and forensic applications since the nineties of the last century. Compared to minisatellites, they are more accurate, faster to sequence, and cost effective [33, 34]. The regions harboring them can be easily isolated using PCR primers and they can be then sequenced using Sanger machines. To increase the power of discrimination, the scientists use multiple loci among the genome, where the probability that two persons share the same multi-loci profile is almost zero.

The CODIS (Combined DNA Index System) database (https://strbase.nist.gov) includes a large number of STR loci and their characterization. It also includes PCR primers that can be used to sequence them. For forensic applications, there are two general standards: The USA standard composed of 20 loci and the European standard composed of 12 loci, and some of these loci's are in both standards [34–36].

The USA standard started with 13 STR loci, and was later expanded in January 2017 to include another set of 7 STR loci. The first core set composed of 13 FBI loci are summarized in Table 1. The extra seven loci are D1S1656, D2S441, D2S1338, D10S1248, D12S391, D19S433, and D22S1045 (not shown in the table). Most of the loci have simple tandem repeat structure composed of multiple units, each of four

Table 1 The core set of 13 STR loci for human identification. The symbol *n* refers to the number of repeated units

Locus		
D3S1358	TCTA$[TCTG]_{n1}[TCTA]_{n2}$	n1 = [1–3], n2[10–15]
D5S818	[AGAT]n	n = [7–15]
D7S820	[GATA]n	n = [6–14]
D8S1179	[TCTA]n1$[TCTG]_{n2}[TCTA]_{n3}$	n1 = [0-2], n2 = [1, 2], n3 = [7–15]
D13S317	$[TATC]_{n1}$ $[AATC]_{n2}$	n1 = [7–15], n2 = [0-1]
D16S539	$[GATA]_n$	n = [5–15]
D18S51	$[AGAA]_n$	n = [8–27]
D21S11	$[TCTA]_{n1}$ $[TCTG]_{n2}$ $[TCTA]_{n3}$ TA $[TCTA]_{n4}$ TCA $[TCTA]_{n5}$ TCCA TA $[TCTA]_{n6}$ TA TCTA	More complex
CSF1PO	[AGAT]n	n = [5–16]
FGA	$[TTTC]_{n1}$ TTTT TT $[CTTT]_{n2}[CTGT]_3[CTTT]_{14}$ $[CTTC]_4[CTTT]_3$CTCC$[TTCC]_4$	More complex
TH01	$[AATG]_{n1}$ATG$[AATG]_{n2}$	n1 = [3–12], n2 = [0–5] up to 12 units; others are more complex
TPOX	$[AATG]_{n1}$ATG$[AATG]_{n2}$	n = [6–14]
vWA	TCTA$[TCTG]_4[TCTA]_{10}$TCCA TCTA	More complex

bases. Some loci have more complex structure, with more than one type of repeated units.

If a locus occurs in a gene region, then they are usually named after the gene itself. The last five loci in Table 1 come from the genes FGA, TH01, TPOX, and vWA. The other loci not in a gene region start with the symbol D. There are a number of commercial kits following either standards or use extra loci [37].

In this chapter, we will give two examples for the use of microsatellites in forensics. The first is for paternity testing and the second is for forensic investigation.

Paternity testing

Figure 3 shows an example of one microsatellite locus in a mother and father. It also shows the combinations any of their children may have. It is important to note that the human genome is diploid, in which there are two copies of the genome: one comes from the mother and the other comes from the father. It is possible that two brothers have the same combination at one locus, but it is almost impossible that they share the same combination on all loci. This is because the recombination process during fertilization of fetus cells which shuffles genetic material at chromosomal regions. Figure 4 shows a real example using a commercial kit over a set of some markers. In this example, one can confirm that the child inherited loci from the parents. If the markers do not lead to enough matching, the result of the paternity test is negative.

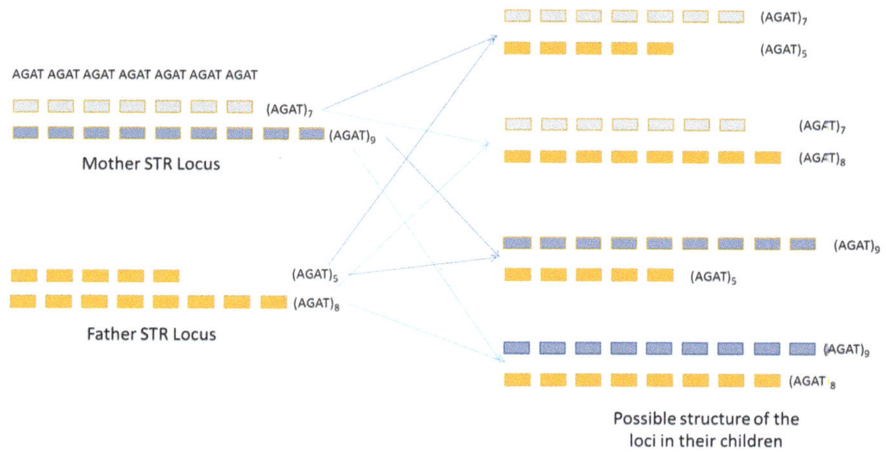

Fig. 3 Example of an STR locus in mother and father. The possible structure of the locus in prospective children is shown

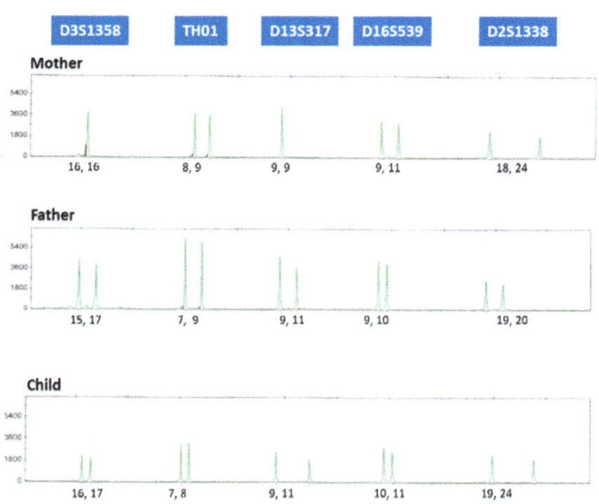

Locus	Mother	Father	Child
D3S1358	16, 16	15, 17	16, 17
TH01	8, 9	7, 9	7, 8
D13S317	9, 9	9, 11	9, 11
D16S539	9 ,11	9, 10	10, 11
D2S1338	18 , 24	19, 20	19, 24

Fig. 4 Example of an STR locus in mother, father, and their child as shown in analytical software when using multi-loci. The table on the right summarizes the length of the microstallite loci for the family members

Forensic investigation

In forensic investigation, the suspect profile at all loci is compared to a database of profiles. The higher the match at all loci, the higher the relatedness. Many authorities worldwide developed a database including loci from suspect persons. This database can help identify criminals who left their DNA in crime scene. Other countries opted to establishing a database for the whole population, to help in identifying missing persons in addition to criminal investigations.

6 Phenotyping Based Techniques

The term phenotype refers to the physical or visible traits. These traits can be categorized into the following:

- *Body traits*, including height, shape, and weight.
- *Pigmentation* (hair, skin and eye color), and
- *Facial features*.

Studying the physical traits of the pea plant in different generations lead to the emergence of the genetic science by Gregor Mendel. Surprisingly, finding the genes responsible for the physical traits and their inheritance model has not been easy for human. Many physical traits in human are more complex and they do not fall into the monogenic inheritance pattern, where single gene (variation) lead to one variation in one trait. In fact, the study of the genetic basis of the human visible traits could make relevant progress only after the automation of the DNA sequencing technology.

Predicting the pigmentation traits reached a good level of accuracy in the last few years. The commercial package HIrisPlex uses DNA variants for predicting the color of eye, hair, and skin [38]. The first version of this package used 24 variants for only eye and hair color. A recent version of this package includes 41 variants and can also be used to predict skin color. The use of these methods have been validated in [39, 40], and in [41]. In the last few months, these methods have been used in some anthropological studies such as the "Cheddar Man" [28], the Jomon Woman in Japan, and the Ava from Caithness [29].

Predicting facial traits is more complex. Identifying the genes and variants (mutations) that affect facial traits was the topic of many papers, notably [42–44].

The company Human Longevity published a prediction system for visible traits [45], as a part of an ambitious project to sequence one million genomes with well characterized phenotypes. The facial prediction module of their system, however, got criticized by a number of researchers who made some re-assessment of the experimental results in the paper [46]. The major critique was that the prediction accuracy was not validated using enough number of cases, and the prediction algorithm could only differentiate/predict faces at the ethnic group level. Specifically, the model in the paper was trained with small number of individuals from 6 ethnic groups, including African, European, Latino, East Asian, and South Asian, and the experiment could only demonstrate that the prediction can assign the query genome to one of the group.

On the commercial side, the company Parabon NanoLabs already provides facial prediction services. The company receives an increasing number of requests for forensics and criminal investigations, when traditional DNA-based identification fails and the suspect DNA is not in the DNA fingerprint database. The algorithms of Parabon are not published yet and it is not known if their prediction is fully automated from DNA information or it includes some human touches.

Detecting the markers for physical traits can be achieved using either NGS or SNP arrays. NGS can either be used for sequencing the whole genome or just some regions of interest. Illumina or Thermo Fischer sequencers are usually used in practice for this purpose. After sequencing, the raw data is analyzed using a bioinformatics pipeline to spot the variations (SNPs and Indels) associated with the phenotype in study. The variations are then analyzed using machine-learning based methods to predict the respective traits. In SNP arrays, genomic segments are not sequenced. With SNP array, one can only decide if there is a variant at certain positions or not. Although it is named after SNPs, SNP arrays can also probe Indels (insertions/deletions) in addition to SNPs (single nucleotide polymorphisms). SNP arrays are used for the traits where the locus of the variant is well known. Modern SNP arrays can include hundred thousands of variants. But focused versions of it including most significant ones in the range of tens or hundreds of variants is more cost effective.

7 Conclusion

DNA-based identification techniques are based on examining certain genomic sites that show variability among individuals. Traditional methods based on PCR and Sanger sequencing usually target short tandem repeats at certain loci and they produce a profile per an individual. To increase the power of discrimination, at least twenty microsatellite loci are recommended by US and European standards. There are two major use case scenarios: First, one compares a given DNA profile (set of markers) of one sample to that of another one. Second, to compare a target DNA profile to a database of profiles.

Recent phenotyping-based methods based on NGS can provide extra information about the person's visible traits, like color, face, and ancestry. This information can be useful when the individual or his relatives are not in any DNA profile database.

More recently, scientists investigated the use of other markers for DNA-based identifications. The markers associated with epigenetics started to prove success. Epigenetic is the study of DNA methylation, which occurs in the cell to modify the structure of RNA without changing DNA. DNA methylation is affected by environmental changes, and the methylation patterns in individual genomes make it possible to estimate the age and to discriminate among identical twins.

Scientists also investigate the use of microbial DNA as a biometric marker. The microbial DNA can come from the environment, such as those that attach to a suspect from crime scene. These can prove or disprove presence at the site. Moreover, the microbiomes within the human body vary among individuals and can also be used as a kind of biometric marker to solve some cases.

The continuous development of DNA sequencing technologies will help achieve more progress in recent research topics. It is expected that the whole genome sequencing cost drops to around 100 USD in the near future. This rising revolution will certainly lead to huge advancement in the prediction of visible traits, the use of epigenetic markers, and the integration of microbial information. It is expected that these methods will be used in large scale studies and will encourage the development of population size genomic databases.

References

1. International Human Genome Sequencing Consortium: Initial sequencing and analysis of the human genome. Nature **409**, 860–921 (2001)
2. Venter J.C., et al.: The sequence of the human genome. Science (80-.) **291**(5507), 1304–1351 (2001)
3. Debrauwere, H., Gendrel, C.G., Lechat, S., Dutreix, M.: Differences and similarities between various tandem repeat sequences: minisatellites and microsatellites. Biochimie **79**(9–10), 577–586 (1997)
4. Ramel, C.: Mini- and microsatellites. Environ. Health Perspect. **105**(Suppl 4), 781–9 (1997)
5. Mirkin, S.M.: Expandable DNA repeats and human disease. Nature **447**(7147), 932–940 (2007)
6. Doi, K., et al.: Rapid detection of expanded short tandem repeats in personal genomics using hybrid sequencing. Bioinformatics **30**(6), 815–22 (2014)
7. Fan, H., Chu, J.-Y.: A brief review of short tandem repeat mutation. Genomics. Proteomics Bioinform. **5**(1), 7–14 (2007)
8. Sanger, F., Coulson, A.R.: A rapid method for determining sequences in DNA by primed synthesis with DNA polymerase. J. Mol. Biol. **94**(3), 441–448 (1975)
9. Bentley, D.R., et al.: Accurate whole human genome sequencing using reversible terminator chemistry. Nature **456**(7218), 53–59 (2008)
10. Merriman, B., I.T. R&D Team, Rothberg, J.M.: Progress in Ion Torrent semiconductor chip based sequencing. Electrophoresis **33**(23), 3397–3417 (2012)
11. Rothberg, J.M., et al.: An integrated semiconductor device enabling non-optical genome sequencing. Nature **475**(7356), 348–352 (2011)
12. Armour, J.A.L., et al.: Minisatellite diversity supports a recent African origin for modern humans. Nat. Genet. **13**(2), 154–160 (1996)
13. Jobling, M.A., Bouzekri, N., Taylor, P.G.: Hypervariable digital DNA codes for human paternal lineages: MVR-PCR at the Y-specific minisatellite, MSY1 (DYF155S1). Hum. Mol. Genet. **7**(4), 643–653 (1998)
14. Gibbs, R.A., et al.: A global reference for human genetic variation. Nature **526**(7571), 68–74 (2015)
15. Durbin, R.M., et al.: A map of human genome variation from population-scale sequencing. Nature **467**(7319), 1061–1073 (2010)
16. Gettings, K.B., et al.: A 50-SNP assay for biogeographic ancestry and phenotype prediction in the U.S. population. Forensic Sci. Int. Genet. **8**(1), 101–108 (2014)
17. Gudbjartsson, D.F., et al.: Large-scale whole-genome sequencing of the Icelandic population. Nat. Genet. **47**(5), 435–444 (2015)
18. Chen, H.: Population genetic studies in the genomic sequencing era. Dong wu xue yan jiu = Zool. Res. **36**(4), 223–32 (2015)
19. Carrasco-Ramiro, F., Peiró-Pastor, R., Aguado, B.: Human genomics projects and precision medicine. Gene Ther. **24**(9), 551–561 (2017)
20. Roewer, L.: DNA fingerprinting in forensics: past, present, future. Investig. Genet. **4**(1), 22 (2013)
21. Saad, R.: Discovery, development, and current applications of DNA identity testing. Bayl. Univ. Med. Cent. Proc. **18**(2), 130–3 (2005)
22. Jeffreys, A.J., Brookfield, J.F.Y., Semeonoff, R.: Positive identification of an immigration test-case using human DNA fingerprints. Nature **317**(6040), 818–819 (1985)
23. Alonso, S., Armour, J.A.: MS205 minisatellite diversity in Basques: evidence for a pre-Neolithic component. Genome Res. **8**(12), 1289–1298 (1998)
24. Rogers, E.J., Shone, A.C., Alonso, S., May, C.A., Armour, J.A.: Integrated analysis of sequence evolution and population history using hypervariable compound haplotypes. Hum. Mol. Genet. **9**(18), 2675–2681 (2000)
25. Brión, M., Cao, R., Salas, A., Lareu, M.V., Carracedo, A.: New method to measure minisatellite variant repeat variation in population genetic studies. Am. J. Hum. Biol. **14**(4), 421–428 (2002)

26. Yuan, Q.-H., et al.: Minisatellite MS_{32} alleles show population specificity among Thai, Chinese, and Japanese. J. Mol. Evol. **68**(2), 126–133 (2009)
27. Foster, E.A., et al.: Jefferson fathered slave's last child. Nature **396**(6706), 27–28 (1998)
28. Brace, S., et al.: Population replacement in early Neolithic Britain. bioRxiv, 267443 (2018)
29. Hoole, M., et al.: 'Ava': a Beaker-associated woman from a cist at Achavanich. Proc. Soc. Antiq. Scotl. **147**, 73–118 (2018)
30. Abouelhoda, M., Giegerich, R., Behzadi, B., Steyaert, J.M.: Alignment of minisatellite maps based on run length encoding scheme. J. Bioinforma. Comput. Biol. **7**(2), 287–308 (2009)
31. Abouelhoda, M., El-Kalioby, M., Giegerich, R.: WAMI: a web server for the analysis of minisatellite maps. BMC Evol. Biol. **10**, 167 (2010)
32. Edwards, A., Civitello, A., Hammond, H.A., Caskey, C.T.: DNA typing and genetic mapping with trimeric and tetrameric tandem repeats. Am. J. Hum. Genet. **49**(4), 746–756 (1991)
33. Coble, M.D., Butler, J.M.: Characterization of new miniSTR loci to aid analysis of degraded DNA. J. Forensic Sci. **50**(1), 43–53 (2005)
34. Butler, J.M.: Forensic DNA Typing : Biology, Technology, and Genetics of STR Markers. Elsevier Academic Press (2005)
35. Butler, J.M.: Genetics and genomics of core short tandem repeat loci used in human identity testing. J. Forensic Sci. **51**(2), 253–265 (2006)
36. Hares, D.R.: Selection and implementation of expanded CODIS core loci in the United States. Forensic Sci. Int. Genet. **17**, 33–34 (2015)
37. Moretti, T.R., et al.: Population data on the expanded CODIS core STR loci for eleven populations of significance for forensic DNA analyses in the United States. Forensic Sci. Int. Genet. **25**, 175–181 (2016)
38. Chaitanya, L., et al.: The HIrisPlex-S system for eye, hair and skin colour prediction from DNA: Introduction and forensic developmental validation. Forensic Sci. Int. Genet. **35**, 123–135 (2018)
39. Walsh, S., et al.: Developmental validation of the IrisPlex system: determination of blue and brown iris colour for forensic intelligence. Forensic Sci. Int. Genet. **5**(5), 464–471 (2011)
40. Pneuman, A. Budimlija, Z.M., Caragine, T., Prinz M., et al.: Verification of eye and skin color predictors in various populations. Leg. Med **14**(2), 78–83 (2012)
41. Walsh, S., et al.: The HIrisPlex system for simultaneous prediction of hair and eye colour from DNA. Forensic Sci. Int. Genet. **7**(1), 98–115 (2013)
42. Shaffer, J.R., et al.: Genome-wide association study reveals multiple loci influencing normal human facial morphology. PLoS Genet. **12**(8), e1006149 (2016)
43. Claes, P., et al.: Modeling 3D facial shape from DNA. PLoS Genet. **10**(3), e1004224 (2014)
44. Claes, P., et al.: Genome-wide mapping of global-to-local genetic effects on human facial shape. Nat. Genet. **50**(3), 414–423 (2018)
45. Lippert, C., et al.: Identification of individuals by trait prediction using whole-genome sequencing data. Proc. Natl. Acad. Sci. U. S. A. **114**(38), 10166–10171 (2017)
46. Erlich, Y.: Major flaws in "Identification of individuals by trait prediction using whole-genome". bioRxiv, 185330 (2017)

A Review on ECG-Based Biometric Authentication Systems

Mohamad O. Diab, Alaa Seif, Maher Sabbah, Mohamad El-Abed and Nijez Aloulou

Abstract The objectives of this chapter is three-folds: First, it presents an overview of the existing ECG benchmarks used for designing ECG-based authentication systems. Second, it presents the literatures of authentication systems that used fiducial and non-fiducial features. Third, it presents a methodology that uses both fiducial and non-fiducial features and several data mining classification techniques for individuals' authentication. Moreover this chapter investigates the pertinent features using a large database of healthy and unhealthy subjects with different heart diseases.

1 Introduction

Biometric authentication solutions are considered promising solutions to improve the traditional authentication systems that are based on what we know (such as passwords) and/or what we have (such as badge). Over the last decade, the Electrocardiogram (ECG) became known as an emerging biometric instrument for individual identification and verification. ECG varies among people due to their diverse heart anatomy. At present, ECG is a popular research topic in the area of physiological biometrics [1]. It is the measurement of the electrical activity of the heart performed or measured on the chest. The ECG biometric has several advantages related to:

- Universality: Among the primary strengths that encompass the practice of ECG signals as biometrics is that the requirement for universality is satisfied. This criterion is met given that the electrocardiogram can be naturally monitored from every subject.
- Permanence: The permanence requirement is also satisfied by ECGs, as the main structure of such signals is invariant over a large period of time. This statement does not imply necessarily that special characteristics of the signals do not get

M. O. Diab (✉) · A. Seif · M. Sabbah · M. El-Abed
Rafik Hariri University, Mechref, P.O. Box 10, 2010 Damour—Chouf, Lebanon
e-mail: diabmo@rhu.edu.lb

N. Aloulou
Faculty of Sciences, Lebanese University, Hadath, Lebanon

A. Nait-ali (ed.), *Hidden Biometrics*, Series in BioEngineering,
https://doi.org/10.1007/978-981-13-0956-4_2

distorted. However, the diacritical waves that compose a heartbeat can be observed and recorded through someone's lifetime. In addition, human heart is very well protected in the body, thus environmental factors cannot have great impact on its activity, as opposed to other biometrics.

- Circumvention: Another substantial advantage in the application of ECGs in biometric systems is their robust nature against the application of falsified credentials. The electrocardiogram waveform is controlled by the autonomic nervous system, therefore by a combination of sympathetic and parasympathetic factors. This suggests that every time instance is relatively different, thus difficult to mimic or reproduce. Furthermore, compared to other biometrics such as fingerprint, if not impossible it is far more difficult to steal someone's ECG hence the systems are naturally more robust to spoof attacks.
- Aliveness Detection: ECG offers natural aliveness detection, being only present in a living subject. With this modality the recognizer can trivially ensure sensor aliveness. Other biometric modalities, such the iris or the fingerprint require additional processing to establish the aliveness of the reading.

However, the disadvantages of employing the ECG for human identification need to be considered. Any biometric security system that is based on that trait has to be invariant to conditions such as physical activity, mental stress or exercise that affect the morphological properties of the ECG waveform.

The objectives of this chapter is three-folds: First, it presents an overview of the existing ECG benchmarks used for designing ECG-based authentication systems. Second, it presents the literatures of authentication systems that used fiducial and non-fiducial features. Third, it presents a methodology that uses both fiducial and non-fiducial features and several data mining classification techniques for individuals' authentication. Moreover this chapter investigates the pertinent features using a large database of healthy and unhealthy subjects with different heart diseases.

The outline of the chapter is defined as follows: In Sect. 2, we present the ECG signal acquisition and its sources of variability. Section 3 presents a comparison between the different available ECG benchmarks and the extracted fiducial and non-fiducial features. An ECG-based authentication methodology that uses fiducial and non-fiducial features and classification techniques is presented in Sect. 4 and experimental results are presented in Sect. 5. Conclusions and perspectives are given in Sect. 6.

2 Electrocardiograph

2.1 Purpose of ECG

ECG is the measurement of small voltage change on the body of the subject. This voltage change is linked with the trigger and recharge of heart muscles during cardiac

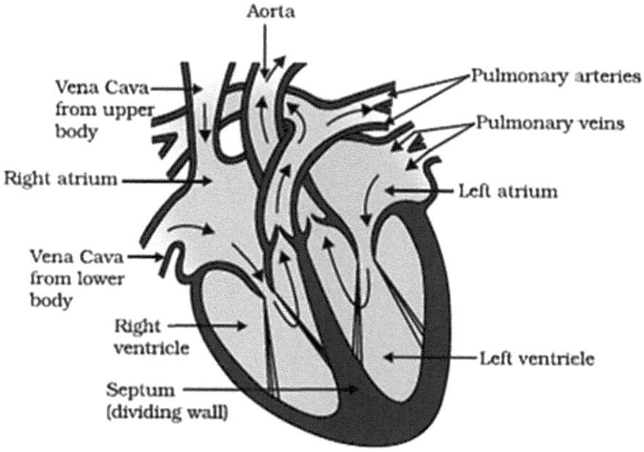

Fig. 1 Schematic of the heart chambers and nodes [3]

activity. In 1924 a Dutch analyst Willem Einthoven first revealed the mechanism of ECG and a Nobel Prize was awarded to him for his work [2].

As shown in Fig. 1, the heart has four chambers. The heart has a natural pacemaker that regulates the pace of the heart. It sits in the upper portion of the right atrium and is a collection of specialized electrical cells known as the sino-atrial (SA) node. Like the spark-plug of an automobile it generates a number of "sparks" per minute. Each "spark" travels across a specialized electrical pathway and stimulates the muscle wall of the four chambers of the heart to contract (and thus empty) in a certain sequence or pattern. The upper chambers or atria are first stimulated. This is followed by a slight delay to allow the two atria to empty. Finally, the two ventricles are electrically stimulated.

As the SA node fires, each electrical impulse travels through the right and left atrium. This electrical activity causes the two upper chambers of the heart to contract. This electrical activity is known as the "P wave" of the ECG. The electrical impulse then moves to an area known as the AV (atrio-ventricular) node. This node sits just above the ventricles. Here, the electrical impulse is held up for a brief period to allow the right and left atrium to continue emptying their blood contents into the two ventricles. This delay is recorded as a "PR interval". Following the delay, the electrical impulse travels through both ventricles. The electrically stimulated ventricles contract and blood is pumped into the pulmonary artery and aorta. This electrical activity is recorded from the surface of the body as a "QRS complex". The ventricles then recover from this electrical stimulation and generate an "ST segment" and T wave on the ECG [4]. As displayed in Fig. 2, the ECG signal is composed of three waves; the P wave, QRS complex and the T wave.

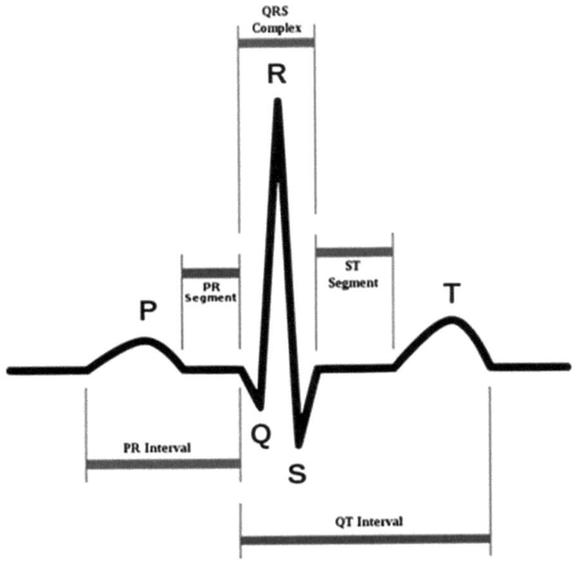

Fig. 2 ECG signal and its P-QRS-T peaks [5]

2.2 *ECG Acquisition*

The ECG signal is usually acquired with a multiple-lead configuration, the most widely applied system is the standard 12-lead ECG. The system is determined by three main sets of lead orientations namely, lead I, II, and III. By convention, lead I measures the potential difference between the two arms. In lead II, one electrode is attached on the left leg and the other one on the right hand. Finally, in lead III configuration, the potential measured is between the left leg and hand. Nevertheless, the ECG can be easily captured with sensor contact of any two points across the heart. As such, the hands and the fingers present a practical solution for real-world applications.

2.3 *Sources of Variability in ECG*

Physiological and geometrical variations of the heart are embedded in the ECG signal. Physiological differences include the heart mass orientation, the conductivity of various areas of the cardiac muscle and the activation order of the heart. On the other hand geometrical differences are due the heart position and orientation relative to the ribs which causes a lot of variance among different people [6]. In other words we can justify these variations with this logic that each person can be identified by a unique code i.e. DNA. In nature the genetic information flows from DNA to RNA (ribonucleic acid) to protein. Eventually protein is responsible for the uniqueness provided by other biometric data (finger print, iris, face, retina, etc.). Therefore, it can be inferred that the uniqueness provided by the ECG signal is inherited from the uniqueness of DNA [7]. Furthermore, other physical factors such as gender, race, and age contribute to the variations of different ECGs [8]. For example, the amplitude of the QRS complex increases from birth to adolescence and then begins to decrease thereafter. It has also been shown that the PR interval increases slightly with increasing age. Sex differences in the ECG signal are more apparent in young adulthood and tend to decrease their effect in other stages. It has been seen that the S wave amplitudes are lower in women than in men between the ages of 18–40 [8].

3 Comparative Study

3.1 ECG Databases

This is an overview of the existing ECG databases used for evaluation of ECG biometrics. ECG is time-dependent and it is naturally affected by the physical and psychological activity of the human body. There is no set benchmark for this application of ECG. Table 1 displays the used ECG databases for biometrics including private and public databases.

3.2 Feature Extraction

3.2.1 Fiducial Features

The majority of the work on ECG biometrics is based on a single channel ECG recording as they were proved sufficient for recognition by Biel et al. [9]. Nonetheless, some worked on recognition using multiple channels [10]. For example, in [11] the identification was based on factor analysis, using PCA, on time, amplitude, area and slope features obtained from a four ECG leads. Results speculated that QRS morphological features are most reliable for personal identification.

Even though most of the previous work is using only ECG recordings [4, 9, 11–28], a few worked on a multimodal system which integrated ECG recognition with either finger prints, face recognition or palm vein [29–31].

Table 1 List of available ECG databases [1]

Database/researcher name	Public/private	Number of subjects	# of sessions	Sensor positions	Conditions or disorders
UofTDB private	Private	1020	Up to 5	Fingers	Sit, stand, exercise, supine and tripod
Zhang and Wei	Private	520	–	Limbs and chest	–
PTBDB	Public	290	Up to 5	Limbs and chest	Healthy and cardiac disorders
Odinaka et al.	Private	269	3	Lower rib cage	Healthy and cardiac disorders
Shen et al.	Private	168		Palms	Sit
QTDB	Public	105		Chest	Exercise and variety of cardiac disorders
LTSTDB	Public	80		Chest	–
EDB	Public	79		Chest	Myocardial ischemia
Wübbeler et al.	Private	74	2–20	Limbs	Supine position
Jang et al.	Private	65	6		Different stress levels
Chan et al.	Private	50	3	Thumbs	–
Agraoti and Hatzinakos	Private	52	Up to 2	Limbs and chest	–
MITDB	Public	47	–	Chest	Arrhythmia
Irvine et al.	Private	43	–	Neck and chest	Different stress levels
Silva et al. and Coutinho et al.	Private	26	–	Chest	Cognitive task
Yao and Wan	Private	20	Up to 4	Limbs	–
Biel et al.	Private	20	4–10	Limbs and chest	Resting
NSRDB	Public	18	1		No significant arrhythmias
Lourenco et al.	Private	16	1	Fingers	–
Homer et al.	Private	12	12 or 18		–
Kim et al.	Private	10		Arms	Resting and exercise

Few papers investigated the physiological state of the subjects during the recording whether they were seated, standing or after exercising [10, 30]. In [30] Poree et al. proved that there is no advantage in comparing shapes recorded in supine rest conditions, as it is classically done, which represents an obvious benefit in biometry. Moreover they proved that there is degradation with time but the results are still acceptable. Most papers didn't look into the effect of patients with cardiac conditions. Instead all the subjects had a healthy cardiac condition. Jekova et al. [11] investigated the effect of medical cardiac conditions (cardiac disease or hypertension) on the ECG biometry. It showed a small decrease in the efficiency for unhealthy subjects.

The majority of papers used either kNN classifier, Nearest center classifier, LDA classifier or Generative model classifier. Instead few papers investigated the use of artificial neural networks or support vector machines.

Process of feature extraction in ECG based biometrics can be separated into two main classes i.e. fiducial and non-fiducial detection approaches. Feature extraction processes that use fiducial techniques refer to the fixed standard of reference in an ECG called biomarkers i.e. three composite waves P, QRS and T for every heartbeat. The fiducial based approach needs computation of amplitude and temporal distances between fiducial points that signify the local maxima and minima, prominent heights, depression between peaks, onsets and offsets of a single heart waveform. Therefore, a fiducial feature completely depends on the precise detection of points of interest. Table 2 summarizes most of the ECG recognition fiducial work done up until now.

3.2.2 Non-fiducial Features

Non-fiducial feature extraction methodologies can be further subcategorized into transform domain approaches and model based approaches [30, 31, 33]. Some algorithms require R-peak detection for segmentation and alignment. Moreover some algorithms require the detection of other characteristic points such as the onset and peak of the P wave, the onset and end of the QRS complex, the peak and end of the T wave [34–36]. Some methods require the detection of all or a subset of the three major components of each heartbeat (P wave, QRS complex, and T wave) for feature extraction [37–39].

On the other hand, others don't use characteristic peaks instead they segment into non-overlapping or overlapping windows and extract features from those windows [39–43]. Table 3 summarizes the previous work using such techniques.

Table 2 Summary of the previous work using fiducial features

Study	Sample size	Condition of subjects	Technique	Classification	Identification performance (%)
Biel et al. [9]	20	–	10 fiducials	GMC	100
Palaniappan and Krishnan [4]	10	– Only healthy	6 fiducials	NN	97.6
Israel et al. [18]	29	–	15 fiducials	LDAC	98
Zhang and Wei [19]	60	– Both healthy and unhealthy	14 fiducials	GMC	82.1
Gahi et al. [20]	16	–	24 fiducials	LDAC	100
Yogendra et al. [21]	25	Both healthy and unhealthy	19 features based on time intervals, amplitudes and angles	Template matching	99
Singh and Gupta [22]	50	– Both healthy and unhealthy	19 fiducials	LMS, HMS	99
Venkatesh and Srinivasan [23]	15	– Only healthy	6 fiducials	KNN	100
Irvine and Israel [24]	29	Only healthy	9 fiducials	GMC	– EER = 0.01
Kyoso and Uchiyama [25]	9	–	4 temporal fiducials	LDAC	>90
Molina et al. [26]	10	–	R-R interval segmentation using morphology getting amplitude and length normalization	Similarity score compared with a threshold	98

(continued)

Table 2 (continued)

Study	Sample size	Condition of subjects	Technique	Classification	Identification performance (%)
Yongjin, et al. [16]	13	Only healthy	15 Temporal features 6 Amplitude features LDA	ED Nearest Neighbor (NN)	98
Fatemian et al. [27]	PTB Database No. of subjects = 294 MIT-BIH Database No. of subjects = 13	Only healthy	QRS detection and delineation of T and P wave then PCA LDA	Template correlation that ranges from 0.75 to 0.98.	99.61–89.27
Khairul et al. [28]	30	Only healthy	With normalized QRS complex	MLP	96.1
Vuksanovic et al. [32]	13	Only healthy	QRS detection, various temporal, amplitude and AR coefficients are extracted	ANN	100
Tiago et al. [15]	63 subjects two sessions separated by 6 months	–	Morphology and segmentation of RR heartbeats	KNN classifier, using the mean wave's Euclidean distances	Test 1 95.2 Test 2 90.2

4 Methodology

4.1 Overview

The Proposed system is trained using labeled ECG recordings. The recordings are processed and features are extracted using Matlab. The extracted features are either fiducial or non-fiducial. The features are tested using Kruskal Wallis statistical test for significance. Finally Machine learning software is used to identify the individuals using their recorded ECG. Figure 3 briefly explains the system employed.

Table 3 A summary of the previous work done using non-fiducial feature sets

Study	Sample size	Condition of subjects	Technique	Classification	Identification performance
Chan et al. [44]	60	Not mentioned	SAECG + WDIST and CCORR	NC	90.8
Saechia et al. [37]	35	Not mentioned	FTF + MLP-BP	NN	97.15
Plataniotis et al. [40]	14	Healthy	AC/DCT + NNC, ED or GLLC	kNN	92.9(NNC) 100(GLLC)
Molina et al. [45]	10	Not mentioned	MSRRS + ED	NC	2(EER)
Wübbeler et al. [46]	74	Healthy	WHVD + NC using 3 channels	NC	98.1
Agrafioti and Hatzinakos [47]	14	Healthy	AC/LDA + NNC, ED using 12 leads	kNN	100
Agrafioti and Hatzinakos [48]	27	Healthy	TMAC + AC/LDA or AC/DCT + NNC, ED (HC)	kNN	96.3(AC/DCT), 100(AC/LDA)
Agrafioti and Hatzinakos [42]	56	Both	APC and PVC Screening + AC/LDA + NNC, ED	kNN	96.42
Chiu et al. [49]	45	Both	DWTMSS + NC, ED	NC	95.71(PHBIA)
Fatemian and Hatzinakos [27]	27	Healthy	MRHB + CC	HMS	99.63
Sufi et al. [38]	15	Healthy	PDM of P, QRS, T waves, NC	NC	100
Wan and Yao [50]	15	Not mentioned	DWTMRRS + MLP-BP	NN	100
Yao and Wan [51]	20	Not mentioned	DWTMRRS + PCA	LMS	91.5
Boumbarov et al. [52]	9	Not mentioned	HMMHS + PCA or LDA + RBFNN	NN	86

(continued)

Table 3 (continued)

Study	Sample size	Condition of subjects	Technique	Classification	Identification performance
Fang and Chan [53]	100	Not mentioned	ECGPSR + SC or MNPD	HMS or LMS	93
Homer et al. [54]	12	Not mentioned	GF and RARMA + NNC, ED	kNN	85.2
Agrafioti and Hatzinakos [55]	52	Healthy	AC/LDA, PT + NNC, ED	kNN	92.3
Coutinho et al. [56]	19	Not mentioned	QHB + ZMCP + MDL	NC	100
Jang et al. [57]	65	Not mentioned	EigenPulse + heartbeat screening	NC	96.92
Li and Narayanan [41]	18	Healthy	HPE + SVMLK	SVM	98.11
Loong et al. [58]	15	Healthy	LPCS + MLP-BP	NN	100
Ye et al. [59]	47	Both	DWT and ICA + SVMRBK (2 CHANNES)	SVM	99.6
Lourenço et al. [35]	16	Not mentioned	MANRHB + NC, ED	NC	94
Tawfik et al. [60]	22	Healthy	DCT of QRS + MLP-BP	NN	99.09
Chataf et al. [33]	10	Healthy	Wavelet networks,	RBF	80–92%

4.2 *Pre-processing*

Figure 4 shows a sample of the ECG signals. The ECG signals are 10 s long segments. Before proceeding to the feature extraction, it is necessary to reduce noise interferences from the ECG signals. Encountered noises are high frequency noise, 50 Hz power line interference, muscle noise and baseline wander. These noises have to be removed or reduced in order for accurate extraction of features from ECG signals. The signal was low-pass filtered to remove 50 Hz power line interference and high frequency interferences using a 3rd order Butterworth digital filter with 3 dB cut-off at 15 Hz. Baseline wander is caused by subject movement or perspiration or excess chest hair. To remove this type of noise, which has a very low spectral range, we first extracted the baseline noise. This was done using a 3rd order low-pass Butterworth

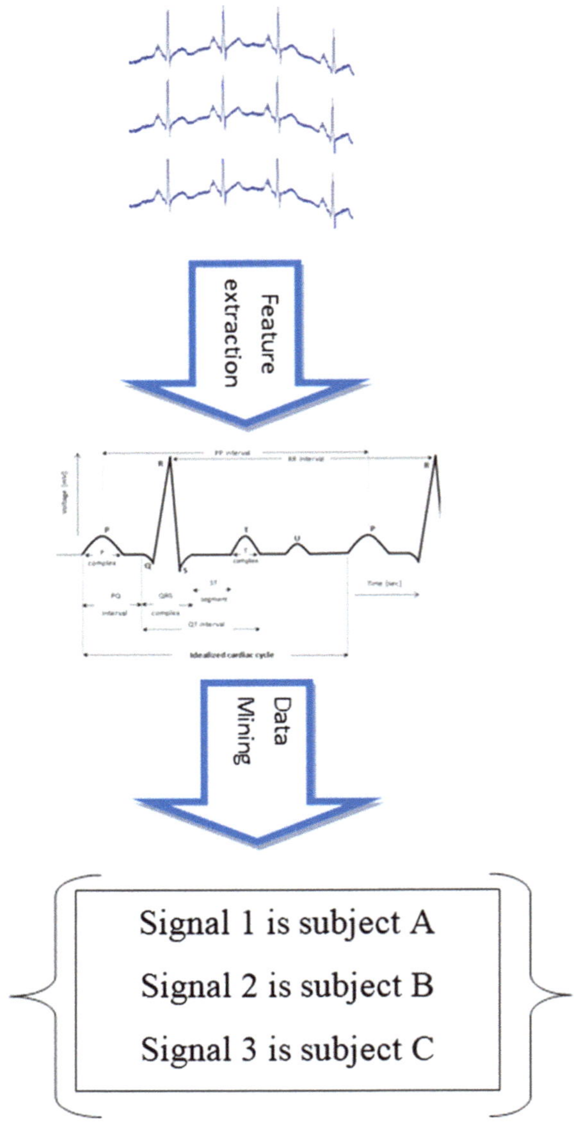

Fig. 3 Schematic of the system employed

digital filter with cut-off at 5 Hz. Once the baseline signal has been detected, the clean signal was obtained by subtracting the baseline signal from the original signal.

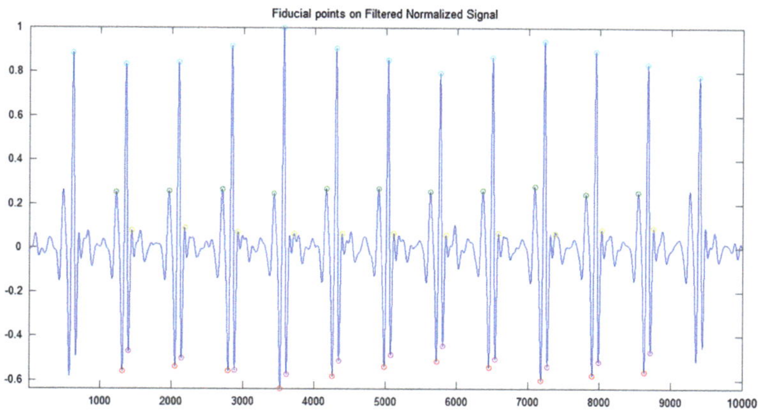

Fig. 4 Filtered ECG recordings

4.3 *Feature Extraction*

4.3.1 Fiducial Features

Pan Tompkin's algorithm was implemented for identification of R peaks. Figure 5 shows the steps of QRS complex detection. After filtration in Fig. (5b) of the signal, the signal is differentiated then squared. After which it a moving window integrator produces a signal. During the derivative stage shown in Fig. (5c), information about the QRS slope is obtained. The squaring displayed in Fig. (5d) intensifies the slope of the frequency response curve of the derivative and reduces the false positive of the T waves. Information about the slope and width of the QRS complex are produced by the moving window integrator. The two sets of thresholds a used are displayed in Fig. (5e). The first thresholds the filtered signal while the second thresholds the signal obtained from the moving window. Figure (5f) shows the pulses of the QRS complex after applying the adaptive threshold on the filtered signal while Fig. (5g) shows the pulse train on the original signal [61].

P-QRS-T peaks were then detected by windowing of the signal around the R-peaks. The Q point was detected as the first downward deflection trough by tracing back 50 ms from R point. Similarly, S point was detected as the first upward deflection peak by tracing forward 50 ms from R point. The features like R amplitude, QR amplitude and RS amplitude were computed. R-R interval was obtained from two consecutive beats. Feature extraction was performed using Matlab.

5 heart beats were used from each recording disregarding the first and last peak to prevent the use of incomplete heart beats. Nine temporal and four amplitude features were extracted into the feature vector. Table 4 summaries the fiducial features extracted from the ECG recordings. Kruskal-Wallis H test was applied to ensure the features showed inter-subject variability and intra-subject stability. The Kruskal-Wallis H test is a rank-based nonparametric test that can be used to determine if there

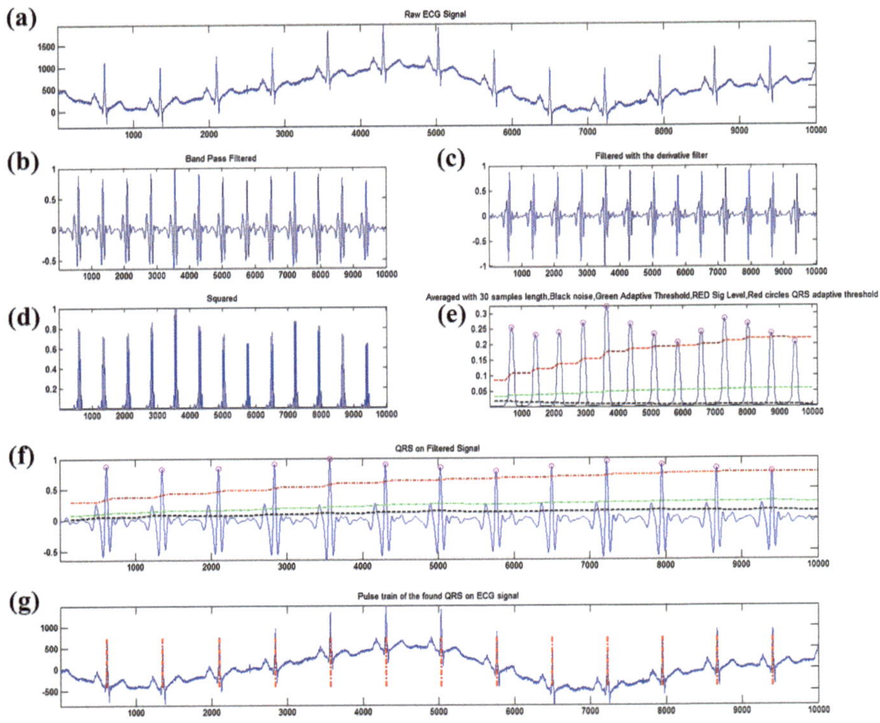

Fig. 5 Descriptions of pre-processing and QRS complex detection using Pan Tompkins algorithm **a** raw ECG signal **b** Band pass filtering of the signal. **c** Filtering with the derivative filter **d** squaring of the signal **e** Peak detection, lowest threshold is noise, middle is adaptive threshold, top is signal level and circles are the R peaks. **f** pulse train on the filtered signal **g** pulse train on the original signal

are statistically significant differences between two or more groups of an independent variable on a continuous or ordinal dependent variable. The Kruskal-Wallis H test was performed on IBM SPSS Statistics.

4.3.2 Non-fiducial Features

The non-fiducial approaches either only detect the QRS peaks or no peaks at all. It could either rely on model based approaches or transform based approaches. The Fourier transform is a useful tool to analyze the frequency components of the signal. However, if we take the Fourier transform over the whole time axis, we cannot tell at what instant a particular frequency rises. Short-time Fourier transform (STFT) uses a sliding window to find spectrogram, which gives the information of both time and frequency. But still another problem exists: The length of window limits the resolution in frequency. Wavelet transform seems to be a solution to the problem

Table 4 Extracted fiducial features

No.	Features	Description
1	r_r	waveform duration between 2 consecutive r-peaks (ms)
2	QRS	QRS waveform duration (ms)
3	RQ	RQ waveform duration (ms)
4	RS	RS waveform duration (ms)
5	RP	RP waveform duration (ms)
6	RT	RT waveform duration (ms)
7	ST	ST waveform duration (ms)
8	PQ	PQ waveform duration (ms)
9	PT	PT waveform duration (ms)
10	RS-amp	RS segment amplitude (μv)
11	RQ-amp	RQ segment amplitude (μv)
12	TS-amp	ST segment amplitude (μv)
13	PQ-amp	PQ segment amplitude (μv)

above. Wavelet transforms are based on small wavelets with limited duration. The translated-version wavelets locate where we concern, whereas the scaled-version wavelets allow us to analyze the signal in different scale. The use of discrete wavelet transform in biomedical signal analysis increased greatly due to its ability to analyze non-stationary signals in frequency domain and time domain. The decomposition of the signal in both domains provides information in low and high frequencies for short and long time intervals. The frequencies in ECG signal vary with time therefore applying Fourier transform does not provide all the information in the signal. The DWT decomposes the signal into details and approximations by cascading of low pass and high pass filters. The approximations are further decomposed. The decomposition level chosen based on the dominant frequency components of the signal and it is limited by the length of the signal. In our work the signal was decomposed to level 6 [62].

There is a variety of mother wavelets that could be used when analyzing a signal as shown in Fig. 6. The two approaches behind the choice of a certain mother wavelet are either theoretically by the similarity of the signal to the mother wavelet or experimentally by testing the performance of different mother wavelets on the signal. The similarity could be quantified using either signal energy or entropy. In this work we looked into mother wavelets that have high energy and/or low entropy with respect to the ECG signal. Then their performance was tested to ensure the obtaining of the highest identification rate. The tested mother wavelets are:

- Daubechies 2 (db2)
- Haar
- Biorthogonal 6.8 (bior6.8)
- Symlets 5 (sym5)
- Coiflets 5 (coif5)

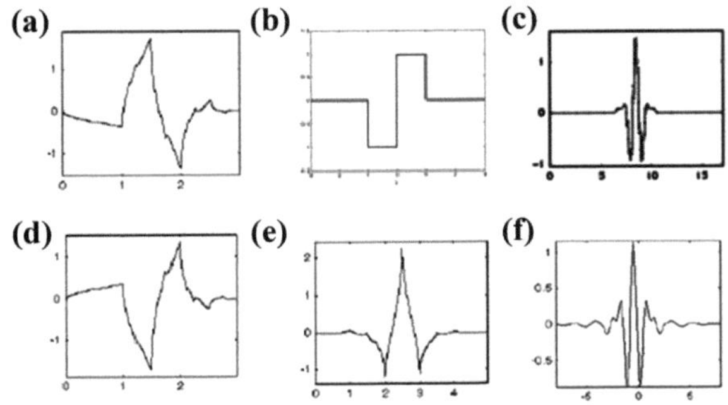

Fig. 6 The mother wavelets used for feature extraction. **a** Daubechies 2 (db2), **b** Haar, **c** Biorthogonal 6.8 (bior6.8), **d** Symlets 5 (sym5), **e** Coiflets 5 (coif5) **f** Discrete Meyer (dmey) [63]

- Discrete Meyer (dmey).

Statistical features over coefficients of the details and the approximation were calculated to decrease the dimensionality of the wavelet feature vector. Those statistical features formed the feature set used for the classification. The following features were used to represent the time frequency distribution of the ECG signal:

- Absolute mean of the coefficients in each sub-band
- Average power of the coefficients in each sub-band
- Standard deviation of the coefficients in each sub-band.

4.4 Feature Selection

For feature selection, Kruskal-Wallis H test (Fig. 7) was applied on feature sets. It ensures that features show inter-subject variability and intra-subject stability. The Kruskal-Wallis H test is a rank-based nonparametric test that can be used to determine if there are statistically significant differences between two or more groups of an independent variable on a continuous or ordinal dependent variable. It is used to test two hypothesis given the null hypothesis that assumes that samples originate from

Fig. 7 Schematic of Kruskal Wallis test

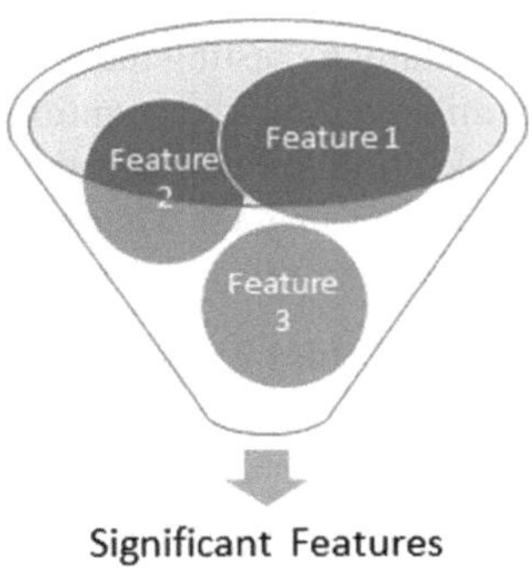

the same population (i.e., equal population means) against the alternative hypothesis which assumes that there is a statistically significant difference between at least two of the subgroups [64]. The Kruskal-Wallis H test was performed on IBM SPSS Statistics.

4.5 Data Mining

4.5.1 Overview

In our work, we attempted to further investigate classifiers that are not the most frequently used in ECG biometrics. Different classification methodologies were used as a comparative study to deduce the most optimum classifier for ECG biometrics. Waikato Environment for Knowledge Analysis (Weka), is a machine learning software developed at the University of Waikato using java. It is open source software that can be downloaded and used for free. Weka contains a command line and a user interface window. It supports tools for data pre-processing, classification, regression, clustering, association rules, and visualization. In our work, we will focus on the classification tool that is used to classify our data. Weka contains many classification algorithms that can be used to approximate the accuracy and visualize different parameter like the ROC curves. As shown in Fig. 8, the applied data mining tools are Neural Networks Multilayer perceptron (NN MLP), Support Vector machine (SVM) using both polykernel and puk kernel, and Decision trees [65].

4.5.2 Artificial Neural Network

Artificial Neural Network is a powerful data mining tool that is inspired from the human nervous system, more precisely the neuron and their connections. Similar to the mind it performs intelligent tasks that require prior knowledge and data learning.

ANNs are built from many different processing elements (PEs). Each PE receives connections from itself and/or other PEs. The interconnectivity defines the topology of the ANN. The signals flowing on the connections are scaled by adjustable param-

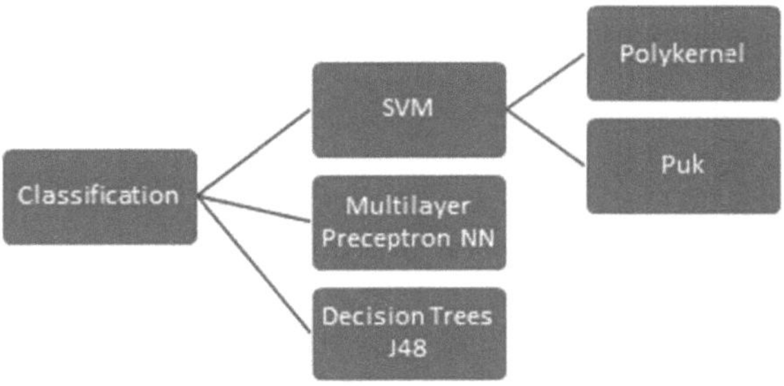

Fig. 8 Tree diagram displaying the classifiers

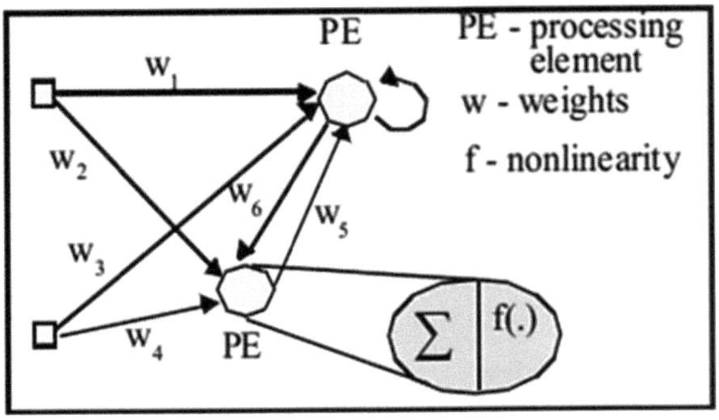

Fig. 9 An artificial neuron network [66]

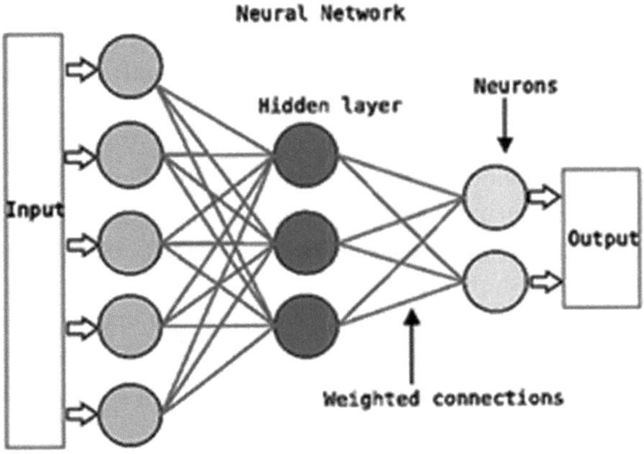

Fig. 10 Schematic of the layers of MLP NN [66]

eters called weights. The PEs sum all these contributions and produce an output that is a nonlinear (static) function of the sum. The PEs' outputs become either system outputs or are sent to the same or other PEs. Figure 9 shows an example of an ANN. Note that a weight is associated with every connection [66].

Multi-layered neural networks are essentially used to deal with data-sets that have a large number of features, especially non-linear ones. Intuitively, the more hidden layers it has, the more 'complex' shapes it can fit. The multilayer perceptron, shown in Fig. 10, consists of three or more layers (an input and an output layer with one or more hidden layers) of nonlinearly-activating nodes and is thus considered a deep neural network. The idea is that each node in the Hidden layer is a function of the nodes in the previous layer, and the Output node is a function of the nodes in the Hidden layer.

4.5.3 Support Vector Machine (SVM)

Support vector machines (SVM) are used for their ability to build linear boundaries between classes by projecting feature vectors in the training set to a high (almost infinite) dimensional space. The SVM was implemented using two different kernels; the polykernel and puk kernel.

SVMs express predictions in terms of a linear combination of kernel functions centered on a subset of the training data, known as support vectors (SV). Assume we have a training set {ti,ci} where ti is the training pattern and ci the label. The training points {ti} are projected into a space H (of possibly infinite dimension) by means of a function $\phi(\cdot)$.

Afterwards it finds an optimal decision hyperplane in this space. For the same training set, different transformations $\phi(\cdot)$ may lead to different decision functions. A transformation is achieved in an implicit manner using a kernel $K(\cdot)$, which can have different forms. Different kernels make use of different parameters leading to difference in performance.

In our work we implemented the SVM using polykernel and Pearson VII universal kernel (PUK). The software used is WEKA which uses the Sequential Minimal Optimization algorithm (SMO) to solve the quadratic programming (QP) problem of SVM. At each time, the SVM will be updated to reproduce new optimal values [64].

4.5.4 Decision Tree (DT)

The third and final classification method is the decision tree C4.5 algorithm. This algorithm is an extension of ID3 algorithm. C4.5 is developed by Ross Quinlan in 1993 and handles continuous and discrete attributes [67].

Assuming there is a training set T and a set of classes {C1, C2, ..., Cn}. The algorithm will perform the decision according the content of T. If T contains samples that belong to the same class Ck, the decision tree for this training set will be a leaf categorizing class Ck. On the other hand, if T contains samples that belongs to different classes, T must be refined into subsets that are aiming to be single class-groups of samples. A test is chosen, based on single attribute, that exclusive outcomes {O1, O2, ..., On}. T will be divided into subsets {T1, T2 ..., Tn} and for each Ti, an outcome Oi of the chosen test will be obtained. T will be a node classifying the test and one branch for each possible outcome. Moreover, C4.5 uses pruning method, the algorithm will remove branches that do not affect the classification and replace them with leaf nodes [64].

5 Experimental Results

The purpose of this thesis is to identify individuals using their ECG recordings. To reach that goal, two experiments were set. The first experiment was done using features extracted from detected peaks. Moreover three different classifiers were tested. While the second experiment does not require peak detection as the extracted features were obtained from the entire recording as is.

The target of the first experiment is to conclude which of the classifiers is most optimal for the application of personal identification using ECG. The second experiment aims to provide a different feature extraction method that does not require peak detection to avoid errors due to wrongly detected peaks.

In Sect. 5.1, we present the first experiment with fiducial features. In Sect. 5.2, we present the second experiment that contains the non-fiducial features.

5.1 *First Approach: Fiducial Features*

In this section we are presenting the results of the first approach that contains the fiducial features. In Sect. 5.1.1, we present the experimental protocol followed by features selection in Sect. 5.1.2. In Sect. 5.1.3, we present the classification results.

5.1.1 Protocol

Thirteen fiducial features are calculated from the five peaks detected in the ECG signal. The experiment is performed on 52 healthy subjects, 173 unhealthy subjects with varied cardiac illnesses and a cumulative test on 239 subjects of hybrid healthy and unhealthy. Examples of healthy and unhealthy ECG recordings are given in Figs. 11 and 12, respectively.

5.1.2 Feature Selection

Kruskal Wallis test is applied after extracting all features. It returns the significant features to be used as input to the classifiers. The interpretation of results should be

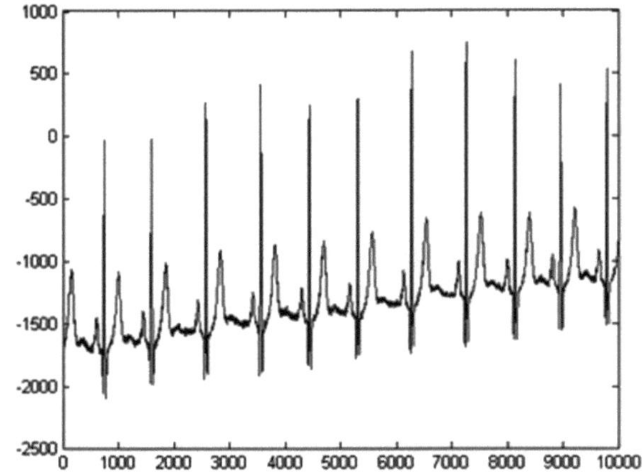

Fig. 11 Example of the healthy ECG recording

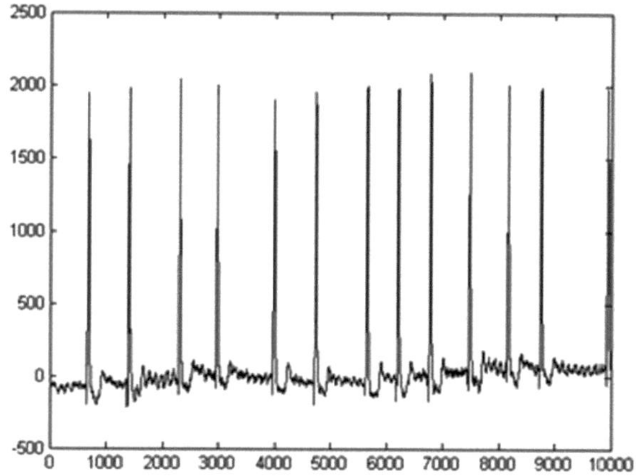

Fig. 12 Example of the unhealthy ECG recording

Table 5 *p*-values of fiducial features

Features	*P*-value
r_r	<0.05
QRS	<0.05
RQ	<0.05
RS	<0.05
RP	<0.05
RT	<0.05
ST	<0.05
PQ	<0.05
PT	<0.05
RS-amp	<0.05
RQ-amp	<0.05
TS-amp	<0.05
PQ-amp	<0.05

done according to the *p*-values. Each feature with a *p*-value less than 0.05 would be significant. The obtained *p*-values for the extracted feature are presented in Table 5. As shown in the table, all the features extracted are significant. Those features were significant because their *p*-value was smaller than 0.05.

5.1.3 Classification

The Fig. 13 displays the results of 52 healthy and 52 unhealthy subjects for comparison of the performance between those two categories, where as Fig. 14 displays the results of a huge database of both categories combined.

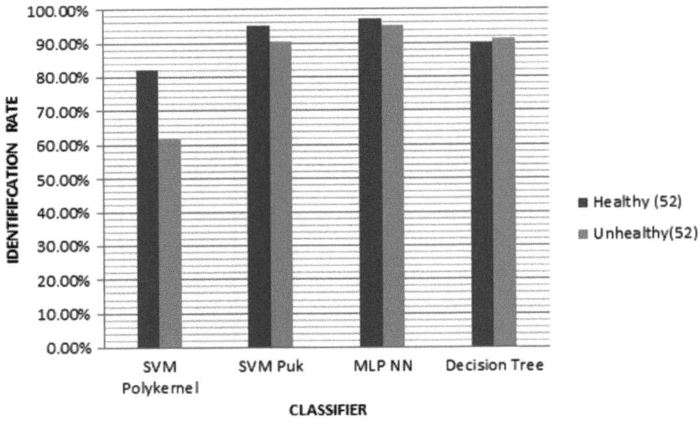

Fig. 13 Correctly classified instances for healthy control versus unhealthy subjects

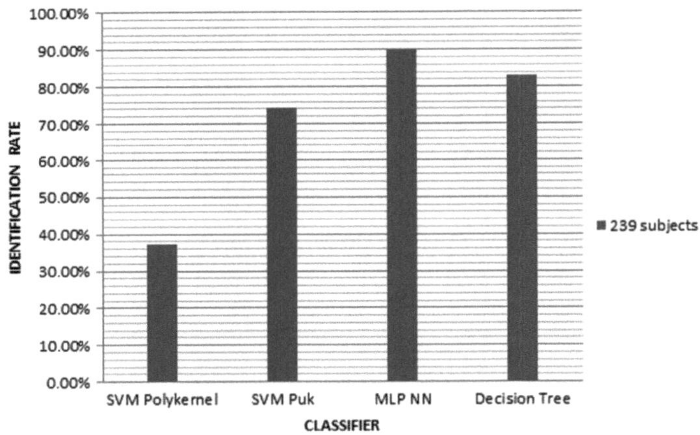

Fig. 14 The identification rate of entire database of 239 subjects

- All classification is performed using 10 cross folds.
- For a sample of same size of healthy and unhealthy subjects, the healthy achieved better performance using all classifiers.
- MLP NN obtained least fall in rate between healthy and unhealthy subjects.
- The 239 subjects are distributed into 52 healthy subjects and 173 unhealthy subjects with varied cardiac conditions.
- The MLP NN obtained the highest performance at 90%.
- There is a huge difference in performance between SVM kernals. The Polykernal obtained 37.3% and puk obtained 74.2%.
- The decision tree came next in performance after MLP NN at 82.9% correctly identified instances.

5.2 *Second Approach: Non-fiducial Features*

In this section we are presenting the results of the second approach that contains the non-fiducial wavelet features. In Sect. 5.2.1, we present the experimental protocol followed by features selection in Sect. 5.2.2. In Sect. 5.2.3, we present the classification results.

5.2.1 Protocol

The DWT decomposed the signal into details and approximations. The decomposition level chosen is based on the dominant frequency components of the signal and it is limited by the length of the signal. It is set to be six levels of decomposition. Six mother wavelets are chosen:

- Daubechies 2 (db2)
- Haar
- Biorthogonal 6.8 (bior6.8)
- Symlets 5 (sym5)
- Coiflets 5 (coif5)
- Discrete Meyer (dmey).

5.2.2 Feature Selection

Kruskal Wallis test is applied after extracting all features. It returns the significant features to be used as input to the classifiers. The interpretation of results should be done according to the p-values. Each feature with a p-value less than 0.05 would be significant. The obtained p-values for the extracted feature are presented in Table 6. The test is performed for statistical features of all the decomposition levels. As shown in the table, all the features extracted are significant. Those features were significant because their p-value was smaller than 0.05.

5.2.3 Classification

The identification performance in this approach is displayed in Fig. 15 and Table 7.

Table 6 p-values of the statistical features of wavelet coefficients

Features	Db2	Haar	Bior6.8	Sym5	Coif5	Dmey
STD (A-D6)	<0.05	<0.05	<0.05	<0.05	<0.05	<0.05
Absolute mean (A-D6)	<0.05	<0.05	<0.05	<0.05	<0.05	<0.05
Band Power (A-D6)	<0.05	<0.05	<0.05	<0.05	<0.05	<0.05

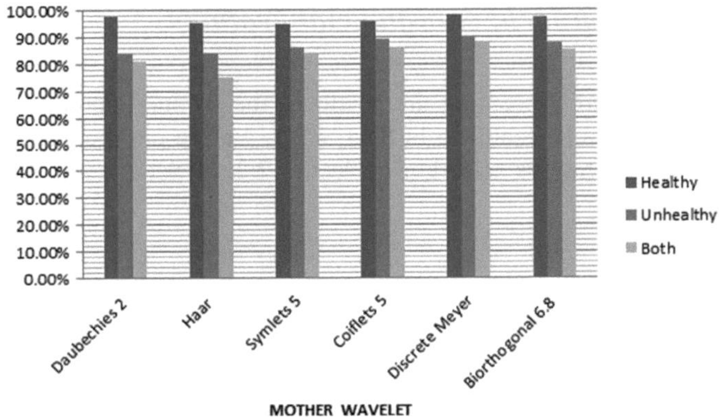

Fig. 15 Graph of identification rate for the six wavelets used

Table 7 Identification rate of the six wavelets

Wavelet	Healthy (%)	Unhealthy (%)	Both (%)
Daubechies 2	97.70	84.45	81.02
Haar	95.50	84.18	75.24
Symlets 5	94.93	86.50	83.61
Coiflets 5	95.63	89.12	86.38
Discrete Meyer	98.60	90.33	87.71
Biorthogonal 6.8	97.20	87.60	85.24

The classifier used in this section is MLP NN since it proved most effective in the first approach.

- The healthy subjects obtained highest performance rate but the decrease between them and unhealthy varied according to the wavelet.
- Discrete Meyer wavelet resulted in highest identification rate in all database categories.
- Coiflets 5 wavelet had the least drop when comparing results between healthy and unhealthy subjects.
- All wavelets had identification rate in the range of 84–98%.
- The wavelet with the worst performance was Haar.

5.3 *Discussion*

This thesis proposed two techniques for individual identification using ECG recordings. It also provides a comparative analysis of the most effective classifier for this application.

The compared feature extraction approaches are 13 fiducial based features obtained from detecting all ECG peaks and statistical features of the wavelet decomposition coefficients. All the presented methodologies obtained high identification performance further confirming the possibility of applying ECG for individual verification and identification. Nonetheless, the non-fiducial based features slightly outperformed the fiducial features. Moreover this report presents a comparison between the possible mother wavelets. Even though all mother wavelets had high identification performance some outranked others with the Discrete Meyer (dmey) obtaining the highest rate 98.6%.

Three different classifiers are studied. The SVM performance fell greatly when the subject number was bigger. Moreover it resulted in a huge gap when using different kernels. Puk is more appropriate for ECG classification and for small samples. In contrary to expected the decision tree's performance did not fall as significantly when the unhealthy subjects were included but it also did not perform as high as the MLP NN. The classification proved MLP NN being most suitable classifier for this task as it gave the highest identification rate on all sample sizes and using both feature sets.

The results show a fall in performance when comparing healthy control to unhealthy subjects. A certain margin of fall is expected between those two categories. A possible solution is to train the classifier of an unhealthy subject with more vectors to overcome the difference. Even though including the unhealthy sample caused a fall in the performance, the resulted rate was satisfactory.

As a conclusion, through this work and previous work ECG biometrics has proven to have great potential.The work done in this thesis further proves that ECG signal could be used for individual identification and verification. It goes to test the theory on a sample that mimics the real life situation.

6 Conclusions and Perspectives

The need for a new biometric system that could be confidential is growing for security reasons. Therefore the ECG signal is extremely fit for this purpose as it is confidential, unique and tests liveliness.

This chapter investigates many aspect of ECG biometrics by using two different approaches. The first approach extracted fiducial features while the second approach extracts non-fiducial features using wavelet transform. Discrete Meyer (dmey) wavelet feature set coupled with the Multi-layer preceptron neural network provided the highest identification rate using a database of healthy (98.6%) and unhealthy subjects (90.33%).

For future perspective, there is a need to form a hybrid feature vector of fiducial and non-fiducial features that is expected to further increase the rate. Moreover the used techniques should be tested on a bigger database. To improve the identification rate of unhealthy subjects the classifier should be trained on more samples. Furthermore, the proposed system is aimed to be implemented in a device for real life identification of individuals.

References

1. Wahabi, S., Pouryayevali, S., Hari, S., Hatzinakos, D.: On evaluating ECG biometric systems: session-dependence and body posture. IEEE Trans. Inf. Forensics Secur. **11**(9), 2002–2013 (2014)
2. Hassan, Z., Omer Gilani, S., Jamil, M.: Review of fiducial and non-fiducial techniques of feature extraction in ECG based biometric systems. Indian J. Sci. Technol. **9**(21), 23–26 (2016)
3. Human Heart Diagram Labeled. Retrieved from http://www.healthsanaz.com/human-heart-diagram-labeled-hhd10.html
4. Sornmo, L., Laguna, P.: Bioelectrical Signal Processing in Cardiac and Neurological Applications, vol. 5, no. 7, pp. 13–15. Academic Press (2005)
5. Potter, L.: Understanding an ECG|Geeky Medics. Retrieved from https://geekymedics.com/understanding-an-ecg/
6. Hoekema, R., Uijen, G.J.H., Van Oosterom, A.: Geometrical aspects of the interindividual variability of multilead ECG recordings. IEEE Trans. Biomed. Eng **48**(51), 551–559 (2001)
7. Suif, F., Khalil, I., Hu, J.: Ecg-based authentication. Handb. Inf. Commun. Secur. **45**(55), 309–331 (2010)
8. Macfarlane, P.W., Lawrie, T.D.V.: The normal electrocardiogram and vectorcardiogram. Compr. Electrocardiol. Theory Pract. Health Dis. **1**(1), 407–457 (1989)
9. Biel, L., Pettersson, O., Philipson, L., Wide, V.: ECG analysis: a new approach in human identification. IEEE Trans. Instrum. Measur. **50**(3), 808–812
10. Poroee, F., Kervio, G., Carrault, G.: ECG biometric analysis in different physiological recording conditions. Signal, Image and Video Process. **10**(2), 267–276 (2014). Springer Verlag
11. Jekova, I., Christov, I., Krasteva, V., Bortolan, G., Matveev, M.: Assessment of the potential of morphological ECG features for person identification. Comput. Cardiol. Conf (CinC), **45**(5), 26–30 (2015)
12. Trader, J.: Expect to see these new biometric modalities become mainstream. Retrieved from http://www.m2sys.com/blog/biometric-hardware/looking-ahead-new-biometric-modalities-horizon/ (2017)
13. Belgacem, N., Bereksi-Reguig, F., Nait-ali, A.: Person identification system based on electrocardiogram signal using LabVIEW. Int. J. Comput. Sci. Engi. (IJCSE) Person **4**(6), 974–981
14. Hassan, Z., Omer Gilani, S., Jamil, M.: Review of fiducial and non-fiducial techniques of feature extraction in ECG based biometric systems. Indian J. Sci. Technol. **9**(21), 850–855 (2016)
15. Tiago, A.: Generic biometry algorithm based on signal morphology information: application in the electrocardiogram signal. Adv. Intell. Syst. Comput. **5**(53), 301–310 (2014)
16. Wang, Y., Agrafioti, F., Hatzinakos, D., Plataniotis, K.: Analysis of human electrocardiogram for biometric recognition. EURASIP J. Adv. Signal Process. **20**(50), 91–95 (2008)
17. Jekov, I., Bortolan, G.: Personal verification/identification via analysis of the peripheral ECG leads: influence of the personal health status on the accuracy. BioMed Res. Int. **215**(5), 1–13 (2015)
18. Israela, S., Irvineb, J.: ECG to identify individuals. Pattern Recognit. Soc. **204**(5), (23–27)
19. Zhang, Z., Wei, D.: A new ECG identification method using Bayes' theorem. TENCON IEEE Region 10 Conf. **20**(23), 22–25 (2006)
20. Gahi, Y., Lamrani, M., Zoglat, A., Guennoun, M., Kapralos, B., El-Khatib, K.: Biometric identification system based on electrocardiogram data. New Technol. Mobility Secur. **15**(20), 56–59 (2008)
21. Singh, Y. Gupta, P.: ECG to individual identification. In: IEEE Second International Conference on Biometrics: Theory, Applications and Systems, vol 8, no. 10, pp. 26–29 (2008)
22. Singh, Y., Gupta, P.: Correlation-based classification of heartbeats for individual identification. Soft. Comput. **15**(3), 449–460 (2009)
23. Jayaraman, S.: Human electrocardiogram for biometrics using DTW and FLDA. 20th Int. Conf. Pattern Recognit. **15**(19), 52–55 (2010)

24. Irvine, J., Israel, S.: A sequential procedure for individual identity verification using ECG. EURASIP J. Adv. Signal Process. **209**(85), 1–14 (2009)
25. Kyoso, M., Uchiyama, A.: Development of an ECG identification system. In: Conference Proceedings of the 23rd Annual International Conference of the IEEE Engineering in Medicine and Biology Society, vol. 23, no. 29, pp. 43–49 (2001)
26. Molina, G., Bruekers, F., Presura, C.: Morphological synthesis of ECG signals for person authentication. Signal Process. Conf. **15**(20), 71–73 (2007)
27. Fatemian, S., Hatzinakos, D.: A new ECG feature extractor for biometric recognition. In: 16th International Conference on Digital Signal Processing, vol. 16, no. 22, pp. 93–95 (2009)
28. Sidek, A., Khalil, I., Smolen, M.: ECG Biometric recognition in different physiological conditions using robust normalized QRS complexes. Comput. Cardiol. (CinC), **52**(55), 120–123 (2012)
29. Singh, Y., Singh, S.: Evaluation of electrocardiogram for biometric authentication. J. Inf. Secur. **3**(1), 39–48 (2012)
30. Tantawi, M., Revett, K., Tolba, A.: A novel feature set for deployment in ECG based biometrics. In: 7th International Conference on Computer Engineering and Systems (ICCES), vol. 7, pp. 186–191 (2012)
31. Coutinho, D.P., Silva, H., Gamboa, H., Fred, A., Figueiredo, M.: Novel fiducial and non-fiducial approaches to electrocardiogram-based biometric systems. IET Biometrics **2**(2), 64 (2013)
32. Vuksanovic, B.: Analysis of human electrocardiogram for biometric recognition using analytic and AR modeling extracted parameters. Int. J. Inf. Electron. Eng. **4**(6), 64–66 (2014)
33. Chantaf, S., Nait-ali, A., Karasinski, P., Khalil, M.A.: ECG modelling using wavelet networks: application to biometrics. Int. J. Biometrics (online) https://doi.org/10.1504/IJBM.2010.033388 (2010)
34. Fatemian, S.Z., Hatzinakos, D.: A new ECG feature extractor for biometric recognition. 16th Int. Conf. Digital Signal Process. **16**(16), 58–59 (2008)
35. Lourenc, A., Silva, H., Fred, A.: Unveiling the biometric potential of finger-based ECG signals. Comput. Intell. Neurosci. **2011**(72), 89–91 (2011)
36. Safie, S., Soraghan, J., Petropoulakis, L.: Electrocardiogram (ECG) biometric authentication using pulse active ratio (PAR). Inf. Forensics Secur. IEEE Trans. **6**(4), 1315–1322 (2011)
37. Saechia, S., Koseeyaporn, J., Wardkein, P.: Human identification system based ECG signal. TENCON 2005 IEEE Region 10, **2005**(1), 1–4 (2005)
38. Sufi, F., Khalil, I., Habib, I.: Polynomial distance measurement for ECG based biometric authentication. Secur. Commun. Netw. Online **2008**(1), 25–28 (2008)
39. Khalil, I., Sufi, F.: Legendre polynomials based biometric authentication using QRS complex of ECG. In: Proceedings of the 4th International Conference on Intelligent Sensors, 4(1), 78–79 (2008)
40. Plataniotis, K.N., Hatzinakos, D., Lee, J.K.M.: ECG biometric recognition without fiducial detection. Proc. Biometric Symp. (BSYM) **2006**(15), 23–26 (2006)
41. Li, M., Narayanan, S.: Robust ECG biometrics by fusing temporal and cepstral information. 20th Int. Conf. Pattern Recogn. (ICPR), **20**(20), 1326–1329 (2010)
42. Agrafioti, F., Hatzinakos, D.: ECG biometric analysis in cardiac irregularity conditions. SIViP **3**(4), 1683–1706 (2008)
43. Agrafioti, F., Bui, F.M., Hatzinakos, D.: Medical biometrics: the perils of ignoring time dependency. Proc. 3rd IEEE Int. Conf. Biometrics: Theory, Appl. Syst. **3**(3), 358–363
44. Chan, A.D.C., Hamdy, M.M., Badre, A., Badee, V.: Wavelet distance measure for person identification using electrocardiograms. IEEE Trans. Instrum. Meas. **57**(2), 248–253 (2008)
45. Molina, G.G., Bruekers, F., Presura, C., Damstra, M., Veen, M.: Morphological synthesis of ECG signals for person authentication. Eur. Signal Process. Conf. **2007**(5), 56–59 (2007)
46. Wübbeler, G., Stavridis, M., Kreiseler, D., Bousseljot, R., Elster, C.: Verification of humans using the electrocardiogram. Pattern Recogn. Lett. **28**(28), 1172–1175 (2007)
47. Agrafioti F., Hatzinakos, D.: Fusion of ECG sources for human identification. In: The 3rd International Symposium on Communications, Control and Signal Processing (ISCCSP 2008), vol. 3, no. 3, pp. 169–171 (2008)

48. Agrafioti F., Hatzinakos, D.: ECG based recognition using second order statistics. In: Proceedings of the Communication Networks and Services Research Conference 2008, vol. 1, no. 1, pp. 222–224 (2008)
49. Chiu, C.-C, Chuang, C.-M., Hsu, C.-Y.: A novel personal identity verification approach using a discrete wavelet transform of the ECG signal. In: The 2nd International Conference on Multimedia and Ubiquitous Engineering, vol. 2, no. 2, pp. 296–299 (2008)
50. Wan, Y., Yao, J.: A neural network to identify human subjects with electrocardiogram signals. Proc. World Congr. Eng. Comput. Sci. **1**(1), 380–382 (2008)
51. Yao, J., Wan, Y.: A wavelet method for biometric identification using wearable ECG sensors. In: Proceedings of the 5th International Workshop on Wearable and Implantable Body Sensor Networks, vol. 5, no. 5, pp. 185–188 (2008)
52. Boumbarov, O., Velchev, Y., Sokolov, S.: ECG personal identification in subspaces using radial basis neural networks. In: IEEE International Workshop on Intelligent Data Acquisition and Advanced Computing Systems: Technology and Applications, vol. 1, no. 1, pp. 446–451 (2009)
53. Fang, S.-C., Chan, H.-L.: Human identification by quantifying similarity and dissimilarity in electrocardiogram phase space. Pattern Recogn. **42**(42), 1824–1831 (2009)
54. Homer, M., Irvine, J., Wendelken, S.: A model-based approach to human identification using ECG. Proc. SPIE **7306**, 45–48 (2009)
55. Agrafioti, F., Hatzinakos, D.: IEEE international conference on signal validation for cardiac biometrics. Acoust. Speech and Signal Proc. (ICASSP) **2010**, 1734–1737 (2010)
56. Coutinho, D.P., Fred, A.L.N., Figueiredo, M.A.T.: One-lead ECG-based personal identification using Ziv-Merhav cross parsing. In: 20th International Conference on Pattern Recognition (ICPR), vol. 20, no. 20, pp. 3858–3861 (2010)
57. Jang, D., Wendelken, S., Irvine, J.M.: Robust human identification using ECG: Eigenpulse revisited. Proc. SPIE **7667**(8898), 196–200 (2010)
58. Loong, J.L.C., Subari, K.S., Besar, R., Abdullah, M.K.: A new approach to ECG biometric systems: a comparative study between LPC and WPD systems. World Acad. Sci., Eng. Technol. **251**(333), 759–764 (2010)
59. Ye, C., Coimbra, M., Kumar, B.: Investigation of human identification using two-lead electrocardiogram ECG signals. Fourth IEEE Int. Conf. Biometrics: Theory Appl. Syst. (BTAS) **4**(4), 1–8 (2010)
60. Tawfik, M.M., Kamal, H.S.T.: Human identification using QT signal and QRS complex of the ECG. Online J. Electron. Electr. Eng. (OJEEE) **3**(1), 383–387 (2011)
61. Pan, J., Tompkins, W.: A real-time QRS detection algorithm. IEEE Trans. Biomed. Eng. **32**(3), 230–236 (1985)
62. Tantawi, M., Revett, K., Salem, A., Tolba, M.: A wavelet feature extraction method for electrocardiogram (ECG)-based biometric recognition. SIViP **9**(6), 1271–1280 (2013)
63. Image Matching. Retrieved from http://old.cescg.org/CESCG-2000/JWalder/index.html
64. Alwan, S., Alabed, M., Zantot, R.: Vehicle recognition from partial images 50–59 (2017)
65. Mitchell, T.: Machine learning. New York, NY: McGraw-Hill (1997)
66. Principe, J., Euliano, N., Lefebvre, W.: Neural and adaptive systems. 591–600 (2000)
67. Data Mining: Practical Machine Learning Tools and Techniques. Retrieved from http://www.cs.waikato.ac.nz/ml/weka/book.html

EEG Biometrics for Person Verification

Bacary Goudiaby, Alice Othmani and Amine Nait-ali

Abstract The purpose of this chapter is to explore the idea of using EEG signals as a biometric modality to recognize individuals. Considered as a variant of Brain Computer Interface (BCI), the concept presented in this chapter deals with a Multi-Channel EEG using Emotiv Epoc system. Mainly, a special interest will be addressed to EEG maps analysis for persons recognition. For this purpose, a generic schema is considered, namely pre-processing, feature extraction, Matching/classification leading to a verification decision.

1 Introduction

Electroencephalogram has been a topic of great interest in the research community since the last century. A discovery by Richard Caton in 1875 had shown that there was fluctuating electrical activity on the surface of the mammalian cerebral cortex (rabbit). A few years later in 1929, Hans Berger, a German physiologist, recorded for the first-time electrical signals from the brain by placing electrodes on the brain [1, 2]. He began to study the electrical activity of the brain in the 1920s and was the first to amplify the signal obtained. In 1937, Berger continues his work on the nature of alpha and beta waves in the human cortex. He also observed that signals from the brain were dependent on the general mental state of the subject (attention, relaxation or sleep) [2].

Berger's work marked the beginning of the use of EEG in clinical and research settings but nowadays, the EEG is used in several other fields. One of the most recent application is Brain Computer Interface (BCI) which is an emerging field for EEG. Nowadays with the development of neuroscience, we know more precisely which areas of the brain are active when we perceive stimuli, when we prepare and execute body movements, or when we learn and memorize things. By having all this information about how the brain works, research has been conducted to develop EEG

B. Goudiaby · A. Othmani · A. Nait-ali (✉)
Université Paris-Est, LISSI, UPEC, 94400 Vitry sur Seine, France
e-mail: naitali@u-pec.fr

A. Nait-ali (ed.), *Hidden Biometrics*, Series in BioEngineering,
https://doi.org/10.1007/978-981-13-0956-4_3

applications. Examples of applications such as controlling a wheelchair, driving a drone, or moving the cursor of a screen.

The activity of the brain can be recorded in the form of a trace called encephalography, and the electroencephalogram is the plot obtained by referring to the etymology of the word. The signals generated are obtained from electrodes placed on the head and near surface of the scalp. These data can be used by a neurologist to analyse and detect possible neurobiological abnormalities. It is also practiced for epileptic patients or those suffering from sleep disorders. In other cases, it can be also used for a diagnosis of a brain death. The brain-dead state is characterized by two EEG recordings showing a separate inactivity of four hours apart [3].

In the field of neurology, the electroencephalogram is used to establish certain diagnoses. It also serves cognitive neuroscience to better understand the brain. An EEG plot makes it possible to identify or characterize psychological states in fundamental neuroscience, or pathological states in neurology.

This examination is an aid to the diagnosis of epilepsy and serves in the follow-up of epileptic patients. It also allows to study sleep disorders and disorders of consciousness and alertness (coma, confusion).

One of the several approaches that have been developed to analyse the EEG signal concern the biometrics field [4], and more specifically hidden biometrics. Why hidden biometrics? Because the EEG signal is not visible and easily accessible as it is the case with some common modalities (e.g. face, hands, etc.). Since the EEG signal can be characterized and modeled, this modality becomes an interesting potential candidate to be used for authentication (verification). In this chapter, a focus on EEG biometric applications is highlighted, including the state of the art of EEG authentication. In addition, a comparative study between the different approaches and methods is provided.

The chapter is organised as follows: we firstly provide a description of EEG characteristics as well as its the acquisition systems. Afterwards an outline of Brain Computer Interface (BCI) is given by providing the different existing systems and devices. In the last section, we will focus on EEG signals for biometrics usages with a processing schema composed mainly of three steps (pre-processing, feature extraction and classification) and for person's verification applications.

2 EEG Acquisition and Characteristics

2.1 Brain Signals Acquisition

Research on how the brain works has helped a lot to understand EEG signals. The brain is a very complex organ and is viewed as as collection of interconnected neurons [5]. Related works to understand the complexity of the human brain are still ongoing but some questions remain unanswered [6]. The brain is composed of two parts called cerebral hemispheres. Each hemisphere is divided into four parts (see Fig. 1), called

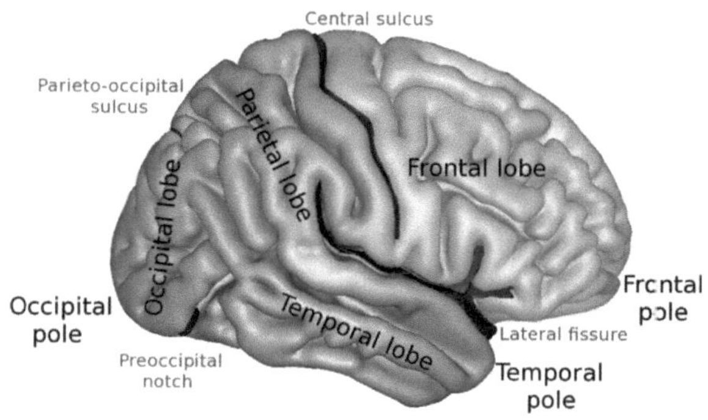

Fig. 1 Cerebral lobes [7]

lobes.

Each lobe has a role to play.

- the frontal lobe, which occupies an important place in the brain, is responsible for resonance, problem solving, movement control, planning, and control of voluntary movements,
- the temporal lobe is involved in the processing of language, memory and the auditory system functions,
- the parietal lobe is intervening in the treatment of sensory information,
- the occipital lobe which allows the recognition of shapes, colors and visual information.

The recorded signals generally have amplitudes that vary from a few microvolts to 100 μV approximately with frequencies ranging from 0.5 to 30–40 Hz [8].

EEG signal is characterized by:

- Amplitude range: wake (Vpp = 100 μV), sleep (Vpp = 300 μV)
- Frequency range: from 0.01 to 100 Hz
- Common EEG artefacts: Eye blinking (EOG), muscle activity (EMG), ambient noise, electrodes movement.

It allows to measure the cerebral electrical activity. The EEG rhythms are separated by frequency range (see Fig. 6). The main five rhythms are:

- Delta rhythms <4 Hz—These are slow waves. They are present when a subject is in a state of deep meditation or a state of sleep without dreams.
- Theta Rhythms (4–7 Hz)—These are waves observed during a deep sleep. These waves play a role in the learning and consolidation of memories.
- Alpha Rhythms (8–13 Hz)—These waves are present during a state of alertness or light meditation. They are associated with the coordination of mental activity and learning.
- Beta Rhythms (14–30 Hz)—These waves are also observed during a waking state when our attention is engaged by cognitive tasks that require making decisions or finding solutions.

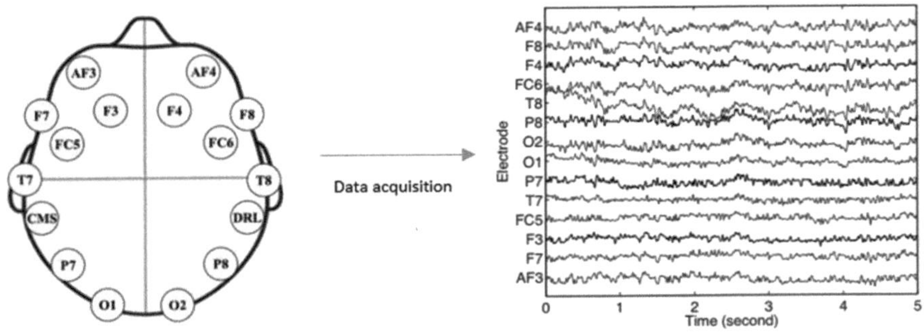

Fig. 2 Example of EEG signal acquisition with electrodes placement. Each waveform represents a signal coming from an electrode. There are 14 electrodes and 2 references CMS and DRL [9]

- Gamma Rhythms >30 Hz—These are the fastest waves. They are associated with information processing by different regions of the brain. They are also present during a state requiring a high level of attention or concentration.

Each rhythm has its own characteristics. In terms of amplitudes, the beta rhythm is mainly observed in the frontal and central regions while the amplitude of the alpha rhythm is the largest in the occipital regions [8] (Fig. 2).

2.2 *Brainwaves Acquisition*

The actions we lead come from the interaction of neurons and chemical reaction of electrical impulses. There is evidence that every part of the brain has a role to play. So, knowing the parts of the brain and their roles, it is then possible to make acquisitions and measurements of impulses from different lobes. There are several methods of acquisition. On one hand, there is the invasive technique for which the electrodes are placed directly on the exposed surface of the brain to record brainwave activity of the subject (e.g. Electrocorticogram) [10] and on the other hand there is the non-invasive techniques for which the brain activity is measured with external sensors. For the EEG acquisition system, the electrodes are placed on scalp based on the international 10–20 EEG placement system (see Fig. 3). The so-called International 10–20 is a well-known method to describe and apply the location of electrodes on the scalp.

The distance between adjacent electrodes are either 10 or 20% of the total front-back or right-left distance of the skull. The nomenclature uses for example, 1–3–5–7–9 for the left hemisphere which represents 10, 20, 30, 40 and 50% of the union to nasion distance respectively.

There are different non-invasive devices that one can use for acquisition.

- **EPOC Emotiv**: it contains 14 channels (AF3, F7, F3, FC5, T7, P7, O1, O2, P8, T8, FC6, F4, F8, AF4). and it has been used for the acquisition of several brain activities (Excitement, Engagement, Relaxation, Interest, Stress, Focus) [11], Fig. 4.

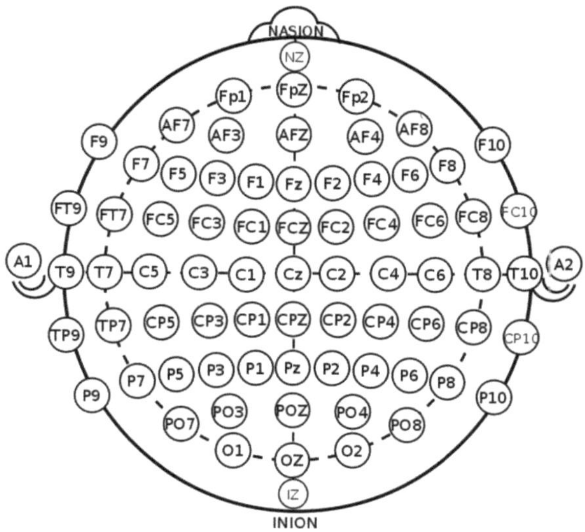

Fig. 3 The international 10–20 EEG placement system F-frontal, T-temporal, C-Central, P-parietal, O-occipital

Fig. 4 Example of Emotiv EPOC headset for EEG signal acquisition

- **Emotiv Insight**: it contains 5 channels (AF3, AF4, T7, T8, Cz) and it has been used for the acquisition of signals generated by brain activity such as Excitement, Engagement, Relaxation, Interest, Stress, Focus [11].
- **Muse**: it contains 4 channels (TP9, AF7, AF8, TP10) and it is designed for the acquisition of signals generated by brain activity such as concentration, relaxation.

Each tool has its own characteristics and has been designed for specific application. Each device also has its own advantages and disadvantages. The non-invasive one has the advantage of not requiring an object implantation in the brain of the subject.

Previously, EEG curves were plotted on graph paper rolls for re-reading by neurologist physicians. Today, this signal is digitally converted to be processed by computer. The signals coming from the brain are analogic and the use of the acquisition devices will transform the acquired analog data into digital data for interpretation and better visualization (see Fig. 5). The recorded data are represented in the form of 1D signals (see Fig. 2) or in 2D form, commonly called "Brain map" (see Fig. 9) which is a schematic representation of the activity of the brain zone showing physical characteristics as a function of the interface used for visualization of brain activity (Fig. 6).

The acquisition process can be realised by using the most common system used in the medical field or a 14-channel referential montage-based commercial EEG device, the Emotiv EPOC which is one of the best low-cost EEG devices in terms of usability. The acquisition process is done in several stages (see Fig. 7).

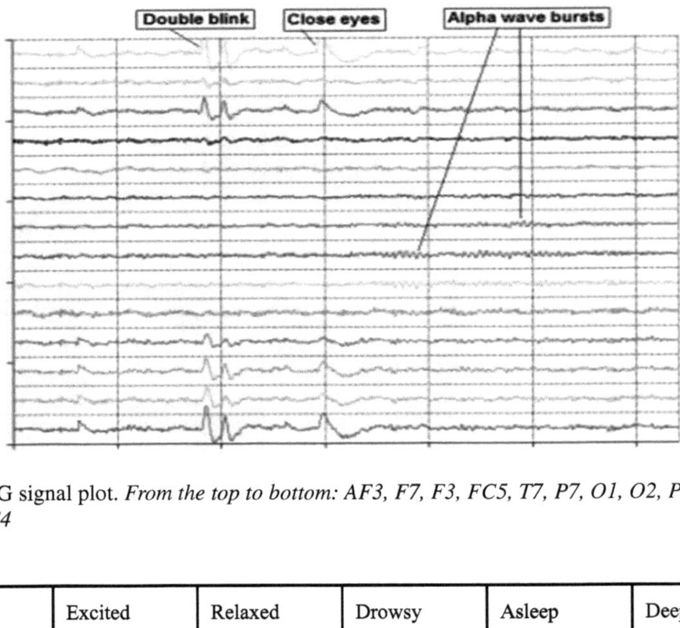

Fig. 5 EEG signal plot. *From the top to bottom: AF3, F7, F3, FC5, T7, P7, O1, O2, P8, T8, FC6, F4, F8, AF4*

Rhythm	Excited	Relaxed	Drowsy	Asleep	Deeply asleep
Waveform	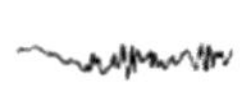	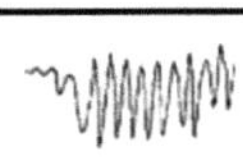	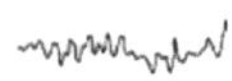		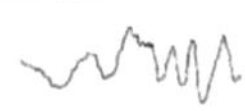

Fig. 6 Rhythms observed in EEG waveform. *When we have high frequency rhythms with low amplitude, this reflects an active brain associated with vigilance or dream sleep. Low frequency and high amplitude rhythms are associated with drowsiness and uncontrolled sleep* [8, 12]

Fig. 7 The EEG acquisition process

2.3 EEG Characteristics

In Table 1, *a* comparison of EEG characteristics with other biometrics modalities is given. Brain activity is neither exposed to the environment nor possible to be captured remotely, so an individual's brain patterns are robust to forgery, face, iris and fingerprints. EEG is more effective against impostor attacks than other biometrics and between different technologies to monitor brain function.

3 Brain Computer Interface

Brain-Computer Interfaces (BCIs) are systems that transform a user's brain activities into messages or commands. BCIs establish a communication channel between the brain and a machine to execute tasks without the use of peripheral nerves and muscles [15, 8]. The BCI system records brain waves and sends them to the computer system to perform the intended tasks interpreted from spontaneous or evoked EEG. Hence the BCI play the role of an interpreter and an executor that perform an action. Its features can help people with paralysis or neuromuscular disorders to execute tasks that require physical movement. The messages conveyed are also used to express an idea or control an object [16].

It was initially designed for biomedical and clinical applications but over time research community has oriented their research on BCI in several other areas. In fact, it is a field of research which tends to develop and extend for various applications [17, 15]. Long remained in the theoretical or fictional field, this concept materialized over time and gains more and more within the researcher's community thanks to devices allowing to make the acquisition and to algorithmic models for a more effective treatment.

BCI is an emerging field that is increasingly developing with studies of brain function. For example, when we want to perform movements that require muscle intervention or when we perform meditation tasks, there are specific brain areas that are active. This is why more and more EEG applications are becoming more efficient using brain activities. This can, for example, help paralyzed patients navigate their wheelchairs [16] or move a cursor on a screen [17].

The goal of the BCI is to be able to record signals from the brain and send them into a machine so that it can interpret them to perform actions. The BCI can therefore

Table 1 Comparison of EEG characteristics with other biometrics modalities [13, 14]

	Univ.	Uniq.	Perm.	Coll.	Accep.	Perf.	Pros and Cons
Fingerprint	• •	• ••	• ••	• •	• •	• ••	Pros: Easy to use/non-intrusive, High accuracy, Long-term stability and ability to enrol multiple fingers, Comparatively low cost Cons: Inability to enrol some users, Affected by skin condition, Sensor may get dirty, Association with forensic applications
Iris	• ••	• ••	• ••	• •	•	• ••	Pros: Potential for high Accuracy, Resistance to impostors, Long term stability, Fast processing cons: Intrusive, High cost
Face	• ••	•	• •	• ••	• ••	•	Pros: Non-intrusive Low cost Ability to operate covertly Cons: Affected by appear-ance/environment, High false non-match rates, Identical twins attack, Potential for privacy abuse
Retina	• ••	• ••	• •	•	•	• ••	Pros: High accuracy, Long-term stability, Fast verification Cons: Difficult to use, Intrusive, Limited applications
Brainwave	• ••	• ••	• ••	• ••	• •	• ••	Pros: non-invasive method is that it does not require an object implantation in the brain of the subject, robust again spoofing attack. Good temporal resolution Cons: Noise on raw data, artefact due to blinks and eye movements [28]

• •• *high*, •• *medium*, • *low*

be considered as an intention interpreter of signals from the brain. That is why work has been done to enable people with disabilities to express themselves and to carry out actions thanks to an actuator. The idea of the BCI can be illustrated by the diagram of Fig. 4.

All intents come from one part of the brain [5]. So, knowing the parts of the brain and their roles, it is then possible to develop diverse BCI applications. In the Fig. 8, the main steps of Brain-Computer Interface system are described. The block diagram of a BCI system consists of four main steps. They include signal acquisition, signal pre-processing, feature extraction, and classification. Signal acquisition involves recording the brain waves and sending them to the pre-processing component to improve the signal and reduce noise and artefacts. The feature extraction step generates the discriminant characteristics for the enhanced signal, reducing the size of the data applied to the classification component. The classification stage consists of translating the produced features into device commands.

Some of the simple actions that are realised, for example with arms or legs are possible to be interpreted with EEG signals recorded from the brain and then executed through a BCI system. When we intend to act, the nervous system produces and carries nervous messages to the organs, mainly the muscles. It is a process called *motor imagery* that is identical to the process that causes a real physical movement. For a BCI system to be able to perform specific actions, a well-structured process with defined steps is required (see Fig. 3). The structure of a BCI includes a system for acquisition and treatment of cerebral signals, a classification system and then translation of these signals into command (take off a drone, write on a screen or wheelchair movement, etc.). To translate these signals into command, during the learning phase, the subject will emit the intention to carry out an action several times. For each intention emitted during learning, a feature set is extracted from the EEG signals and then submitted to a classifier to translate the motor imagery of the

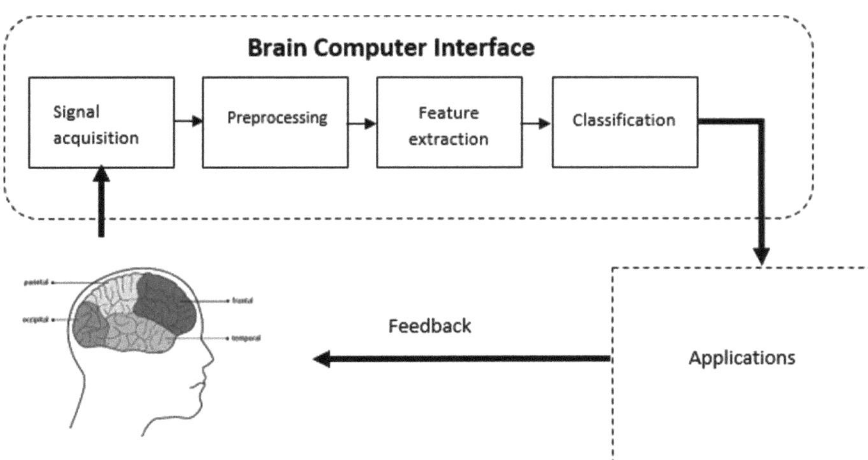

Fig. 8 Block diagram of a brain computer interface

subject into commands. The commands are defined according to the application that will be implemented. In recent work, several BCI applications have been proposed such as brain typing system [17]. This approach was used in [17], the aim of this work is to design an online brain typing system to convert users' intentions into texts and the results obtained gave an accuracy of 95.93 and 94.27% respectively on a public dataset and local dataset acquired using the RNN and CNN models.

4 EEG Biometrics

There are several types of biometric data that can be exploited such as face, voice, fingerprint, palmprint [13] and some hidden biometric data from dental radiography, MRI brain imaging, full body X-ray scanner and Hand X-ray images [18]. One of the main issues in biometrics is spoofing [19]. In the past several solutions have been proposed such as multimodal. There are many related works in this domain. However, it has been shown that some multimodal biometric systems are not robust against spoofing attacks [19].

The EEG modality, like other biometrics modalities, has it own characteristics, its own advantages but also its own disadvantages.

A biometrics system must satisfy a certain number of criteria, such as:

- **Universality** (the chosen biometric modality must be present in all human beings and must be satisfactory for a larger group),
- **Uniqueness** (the biometric characters used for identification or authentication must be unique),
- **Permanence** (the biometric character used must be stable over a period of time. It must not be subject to significant variations),
- **Collectability** (the subject acquisition process should be easy to use and collect easily),
- **Performance** (the biometric system must achieve the desired accuracy for which it was designed),
- **Acceptability** (the biometric chosen for a specific application must be very acceptable to the user. Some biometric characters are more easily accepted than others).

4.1 EEG Biometrics Versus Other Modalities

In this section a brief two different comparison. A comparison EEG characteristic with other common biometrics modalities with the advantages and disadvantages in Table 1. However, there are also emerging hidden biometrics modalities that need to be explored in Table 2, a comparison between EEG and the other Hidden biometrics modalities are given.

Table 2 Comparison between hidden biometrics modalities

Hidden biometrics modalities	Pros	Cons	Characteristics
Dental	Dental features are considered the best suitable for postmortem identification	Requires ante-mortem radiograph Not suitable for non-cooperative subjects	Physiological
ECG	In certain case ECG can be used for continuous authentication	Vulnerability because of change of body or conditions e.g. effects of disease, stress, changing of heartbeat Not suitable for non-cooperative subjects	Physiological
EoG	High recognition rate Intrinsic physical characteristic	The variation of the blink latency may affect the performance; Not suitable for non-cooperative subjects	Behavioral
EEG	High recognition rate, very secure, non-invasive acquisition system, less expensive acquisition method	Noise on raw data, artefact due to blinks and eye movements; Not suitable for non-cooperative subjects	Behavioral

4.1.1 EEG Biometrics Compared to Common Modalities

See Table 1.

4.1.2 EEG Biometrics Compared to Hidden Biometrics Modalities

See Table 2.

4.2 EEG Signal Processing for Biometric Applications: A Review

In the following, the main steps of EEG applications shown in Fig. 9 are detailed.

4.2.1 Pre-processing

The analysis of brain activities plays an important role in neuroscience. The cerebral signals carrying information allow us to authenticate a person. Today it is possible, thanks to the acquisition tools and with its non-invasive character and its high temporal resolution, the signal EEG has many advantages but also has some defects. EEG signals are complex and difficult to identify due to the fact that the signals are likely to be contaminated by intrinsic and extrinsic factors, which does not always facilitate the analysis or use of the latter. Generally, the causes of these noises are multiple [20]. Acquiring signals from the scalp contains sources of noise that can cause problems when analysing, displaying, and restoring EEG signals. Noise in the signal produces errors that lead to a misinterpretation of cerebral signals. Hence the need for pre-treatment. In the simplest cases, conventional artefact removal methods are used such as lowpass, bandpass, or highpass. But these methods are not always effective in the case where the frequency of the signal and the noise overlap [21]. Alternative techniques have been proposed to provide an answer to this problem, such as adaptive filtering [22] and the wavelet transform [23].

Blinking eye is one of the sources of artefact presence in EEG signals. It produces large electrical potential around the eyes known as EOG (Electrooculogram). Then the EEG signal may overlap with artefacts, and one of the solutions proposed by [22] is to use a cascade of three adaptive filters based on a least mean squares (LMS). The ocular artifacts from EEG can be removed using adaptive thresholding of wavelet coefficients [24] while the artefacts related to head movement are removed using an accelerometer placed on the head and independent component analysis [25]. Several solutions have also been proposed. One of the proposed solutions is wavelet

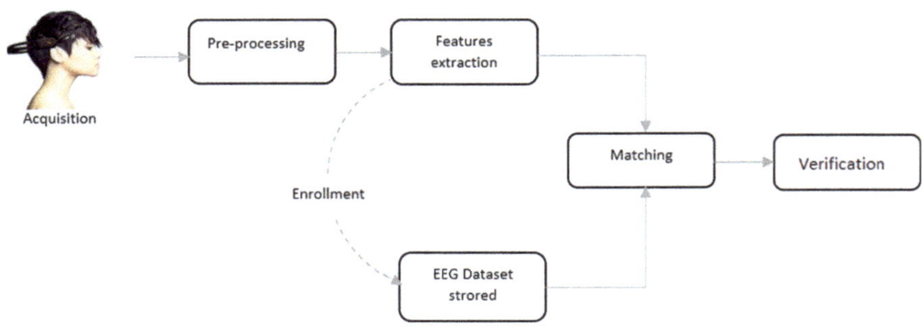

Fig. 9 Block diagram of EEG for person verification

transformed to preserve the main information at best. Solutions have demonstrated the benefits of using wavelet processing to cancel eye blinks without an EOG reference channel [23]. The wavelet transform was also used by Ngoc et al. design filters to remove power lines and baseline noise from acquired signals by identifying the characteristics of recorded EEG signals, the authors performed the wavelet transformation and a threshold method is applied to detect and remove artefacts in EEG signals with flashing eyes [26]. One method has also been used to detect the blink zone based on Wavelet, which allows us to keep the necessary information. Different approaches are based on wavelet transform for ocular artefact removing consisted in using adaptive filter [22, 23]. Reddy et al. proposed a novel method of De-noising EOG signals using Dual Tree complex wavelet transform (DT-CWT). However, the DT-CWT method is best suitable for real time EOG based applications like human-machine communication devices for disabled persons, eye movement analysis and gaming applications [27].

The artefact from EEG could also come from interference with other tools that emit in the same frequency bands or from a limb in motion, which can alter our recorded signal and therefore a pre-processing step is needed [28]. The environment also plays an important role during the acquisition so that the concentration of the subject depends on it. There are different factors to consider when acquiring EEG signals such as ambient noise, visual noise or background music [29]. Research has also focused on the influence of music. Zhou et al. explored the influence of background music on the use of BCIs [30]. In [30], they observed that different performance measures reveal no significant difference when comparing background music and lack of background. In this case, the data could be influenced by the effects of its environment therefore the results obtained could be considered to be false.

Several methods to overcome problems on undesirable signals e.g. artefacts have been proposed by the research community. There are lot of solutions that have been proposed and we can see in [31], the authors gave a review of the most frequently used EEG signal processing techniques. In [20], the author reviewed pre-processing techniques such as the basic filtering of removing certain unwanted frequency bands, AF which is used to attenuate the frequency bands containing the most artefacts and used the Blind Source Separation for artefact removal. The pre-processing task could also depend on the application to be implemented, for example, in [15] where the goal was not to reduce the artefacts but rather to filter by selecting a frequency band between 8 and 12 Hz.

4.2.2 Feature Extraction

The acquisition of EEG signal allows initially to obtain raw data. These data need to be processed, then extract feature before the classification. The step of extracting the feature is very important. There are different methods for extracting features. Each method of feature extraction has it own particularity (see Table 3) and can be classified into three types: time-domain, frequency-domain, and time-frequency domain feature extraction. To authenticate or identify people most effectively several models

Table 3 Description of the advantages and disadvantages of several feature extraction methods using EEG signal

	Pros	Cons
PCA (Principal component analysis)	Dimension reduction without losing data	Unable to process complicated set of data
ICA (Independent component analysis)	Computationally efficient, high performance for large data sized	Require more computational for decomposition
WT (Wavelet transform)	Suitable for non-stationary signals, able to analyse signal in time and frequency domain	Lack of specific methodology to apply to pervasive noise
AR (Auto regressive)	Require short duration of data, reduce spectra loss problems and give better frequency resolution	Not applicable to stationary signals

have been proposed such as Auto-regressive coefficients (AR), Spectral Power (SP), Inter-Hemispheric Power Difference (IHPD), Interhemispheric Linear Complexity (IHLC) [32], Common Spatial Patterns (CSP) [33], Wavelet Transform (WT) [27]. The wavelet transform is often used for pre-treatment [22, 34] but it is also a technique that can be used for the feature extraction indices and even for EEG classification. This is the case in [35] where the reduction as well as the pre-processing of the signals are done with the wavelet transformation.

The extracted features are different according to the acquired data and the designed application. Palaniappan [32] proposed a multiple mental thought identification modal where an Electro-Cap elastic electrode cap was used to record EEG signals from positions C3, C4, P3, P4, O1 and O2 defined by the 10–20 system of electrode placement. In this work the captured data features were extracted using AR modelling. However, Altahat et al. [36] uses Power Spectral Density (PSD) instead of autoregressive (AR) models very often chosen for greater accuracy. Thomas and Vinod [37] take advantage of individual alpha frequency and delta band signals to compose a specific characteristic vector. In this work PSD was preferred but only perform extraction on a gamma band.

4.2.3 EEG Classification

Various research studies have established the feasibility of using EEG signals to classify data for verification purposes. Different approaches of signal processing, feature extraction, and also classification have been proposed with a focus on performance (accuracy) using single-channel or multi-channel EEG signals.

There are a multitude of algorithms that are used to classify and analyse EEG signals. Classification algorithms can be classified into two groups (supervised or unsupervised) depending on the application that will be implemented.

The classification is a very important task for biometrics systems. The extracted features would be used as input for the classification step. In the case of applications using EEG data, several approaches have been proposed such as LDA, kNN, SVM, etc. Palaniappan achieved 100% accuracy by classifying 5 subjects using a linear discriminant classifier [32], as well as a zero false acceptance rate (FAR) and zero negative rejection rate (FRR) using a two-level authentication process [32] while Poulos et al. uses an artificial neural network to classify 4 subjects according to their EEG signals [38].

4.3 *EEG Biometrics: Protocoles (Previous Application of EEG in Biometrics)*

The EEG signal could be considered as a kind of "cerebral fingerprint" that can also be used as a means of authentication. In this section, a study about the existing state-of-the-art methods of EEG based user authentication schemas is given. However, different approaches for authentication based on EEG signals are proposed. In fact, acquisition protocols, feature extraction and classification methods are different depending on the application for which the system is intended.

Person authentication is the process by which the system assures the legitimacy of a person's request for access. Nowadays, authentication through computers has become a daily problem. There are other biometric modalities such as iris, face, retina, and cerebral signals (see Table 1) that have been proposed. EEG signals are increasingly being studied for biometrics. However, in [37], the system is designed so that the user will read numbers that are displayed on a screen while recording the corresponding EEG signals. The captured signals are used to set up the system. This approach focuses primarily on the combined alpha and beta frequency band of 8–30 Hz across all EEG channels, as this is the most appropriate band for EEG signals in Brain Computer Interfaces (BCIs). Common spatial pattern (CSP) values were used as the main characteristics to drive the model. Linear discriminant analysis (LDA) was used as a classification algorithm for a given set of user data with 12 subjects whose acquisition is done on Alpha and Beta rhythms. The approaches may also be different on the choice of the number of electrodes placed on the scalp. It can be seen in [33] where 14 electrodes were used whereas in [37] 2 electrodes were used but the performance in terms of accuracy is the same.

When dealing with the authentication process, three main categories can be considered, such as: password-based, token-based and biometrics-based [39]. Each technique has its own advantages and disadvantages. The feasibility of authentication method using brainwave has been considered in [40] where during the experimentation the purpose was to ask 32 adults' participants to read silently a text. In another study, the feasibility is also based on its inimitable characteristics and uniqueness [9]. Pham et al. showed in their work that an EEG based authentication system has the combined advantages of both passwords-based and biometric-based authentication systems without the drawbacks of these ones [41].

Other approaches have been used to enhance the reliability like multimodal approach [42]. The aim of this approach is to to combine EEG modality with other modality to increase the performances. Blinking of the eyes is often considered as a source of artefacts for EEG signals [31], however it could improve the performance of EEG-based biometric authentication [43]. In [43], a new approach of authentication using EEG signals and EOG signals has been proposed with a database of 31 subjects collected with Neursky Mindwave headphone. The subjects are asked to perform three different tasks of relaxation, visual stimulation and blinking. The same approach has also been practised in a context of continuous authentication [44] where the application is designed for trusted autonomous system. In this approach, a multi-modal biometrics system that continuously verifies the presence of a logged-in user is implemented. Two types of biometric data are used, face images and Electroencephalography (EEG) signals. Both in [44] and [42], they get a significantly good result.

Unimodal and multimodal [44, 42] authentication methods using EEG signals gave good performance. However, it has been shown that the combination of the two-step authentication method with the characteristics of the EEG and thought activities has good potential as a biometric method [35]. The proposed method was tested on five subjects using some reflection activities, a good result is obtained using wavelet coefficients.

Current systems using conventional biometric data such as fingerprints, facial recognition, iris, and retina may be at risk of identity theft and have some disadvantages [42, 45]. Zhang et al. proposed a solution using a multimodal biometric authentication system called DeepKey [46] for better protection against these risks. The goal of their work has been to use two different biometric modalities that are the gait and EEG signal. In their work, two model have been proposed. The first model uses a one-class SVM algorithm and a deep learning model based on a recurrent neural network. The second model combines autoregressive coefficients, an RNN structure and an SVM classifier. The experimental results achieved an overall accuracy of 0.983 with an overall false acceptance rate of 0.0 and a negative rejection rate of 0.019 with a gait data set of 160,000 samples and an EEG dataset of 108,000 samples. To improve performance of authentication, new features have been proposed in [12]. An open-access EEG database containing 109 EEG data signal is used. From the EEG signals, 10 single-channel features (seven spectral and three non-linear) were extracted by performing spectral and non-linear analysis. A distance-based classifier was used to classify the extracted characteristics. They got an equal error performance (EER) of 1%.

In other application, low cost and good performance are considered on designed authentication based in EEG signal [47]. During data acquisition, the subject was asked to perform four mental imaging tasks consisting of baseline measurement, repository member movement, counting, and rotation for 150 s each. The 150 s of data is divided into a second segment, from which the characteristics are obtained. Three features extracted from each electrode: 6th order autoregressive (AR) coefficients, power spectral density and total power in five frequency bands. Two additional sets of features are extracted from interhemispheric data: interhemispheric power differences and interhemispheric linear complexity.

The purpose of authentication is to secure access to a place, access to a multimedia or digital platform. The signals being also biometric data allowed to carry out work on the verification of person based on the EEG. In recent related work, an approach has been proposed for border security applications with EEG biometrics. Currently there are several traditional biometric identification methods used for authenticating people in border security or humanitarian aid activities for example. Face or fingerprint are widely used. EEG signals can also be used to identify both the actual intentions of the applicants by analyzing emotions and identities by biometric analysis [48]. Nakamura et al. has introduced a proof-of-concept for feasible, collectable and reproducible EEG biometrics in the community with an unobtrusive, discret, and convenient to use in-ear EEG device. A robust PSD (Power Spectral Density) and AR (Autoregressive) features has been employed to identify an individual. They performed classification rigorously, without mixing the training and validation data from the same recording days [49]. Poulos et al. showed that lowest person verification error rate could be achieved when each person performed his/her best motor imagery task. In their experiments, they have used Gaussian mixture model (GMM) and support vector data description (SVDD) methods were used for modelling persons. Based on their acquisition protocol, two tasks in the experiments were performed. In the first task, the same motor imagery task of left hand or right hand was applied to all persons. In the second task, only the best motor imagery task for each person was performed.

In the recent years, the research community has interest on EEG based authentication. Several approaches have been proposed with different methods (see Table 4).

5 Study Case

5.1 *Protocol of Data Collection*

This session explains the protocol applied during data collection sessions. Two datasets have been recorded in this project. The goal of this project is to study and propose a scheme that would allow a subject to be authenticated with EEG signals. A small-scale experiment involving 10 participants first to learn how to proceed before conducting a larger scale experiment is realised. The procedure is described in Fig. 10.

In this experiment, the participants were informed about the experimental procedure and made images of the motor before starting the experimental recording. Participants will imagine doing a task as a type of mental activity that involves thinking of a movement without performing it and physically moving their limbs.

Table 4 Summary of related works on EEG signal-based authentication

No.	Number of electrodes	Feature extraction method/features extracted	Classification method	Tasks/number of classes	Number of participants	Reported accuracy/performance
[46]	3	Event-related potentials (ERP)	Normalized cross-correlation classifier	Visual stimulation of 400 images	50	Acceptance rate (FAR) of 0.0 and a False rejection rate (FRR) of 0.019
[34]	14	Common spatial patterns (CSP)	LDA	Imagining four-digit number as cognitive task	12	97%
[32]	6	Auto-regressive (AR), Spectral power (SP), Inter-hemispheric Power difference (IHPD), Inter-hemispheric linear complexity (IHLC)	LDA	Relaxation, math activity, geometric figure rotation, mental letter composition, visual countings	6	100%
[33]	14	Common spatial patterns (CSP)	LDA	Visualize a number while corresponding EEG signals are captured	12	Maximum accuracy recorded with the existing data was 96.97%. Overall system accuracy is 66.67%

(continued)

Table 4 (continued)

No.	Number of electrodes	Feature extraction method/features extracted	Classification method	Tasks/number of classes	Number of participants	Reported accuracy/performance
[45]	1	Two dimensional signals (Alpha wave and Beta wave) recorded and compress it into a one-dimensional signal. For each frequency component, the median magnitude corresponding to that frequency component is calculated over all time at the end to have a one-dimensional vector	kNN	Breathing, simulated finger movement, sport activity, singing/passage recitation, audio listing, color identification, and pass-though	15	99%
[50]	64 and 10	A bivariate measure, Magnitude Squared Coherence (MSC)	KNN	EEG signals are taken from online available Physionet database, where eye opened data on resting state for 1 min for 109 healthy subjects are used, each signal is obtained at 160 Hz	108	100%

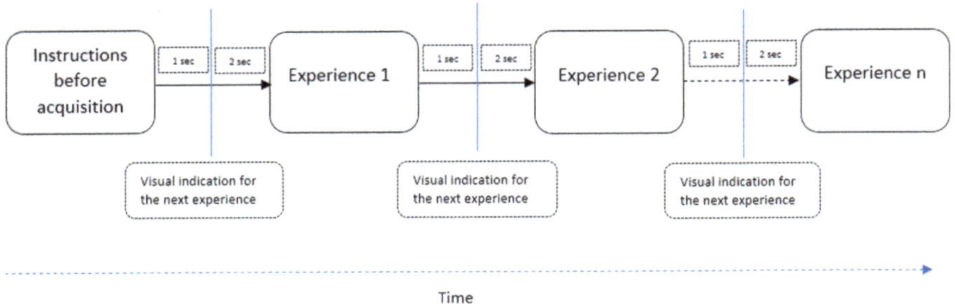

Fig. 10 Protocol of EEG acquisition of this case study. *Inter trial interval of 1 and 2 s interval for a visual indication for action to be done. The record of each experience will take 3 s*

All participants were invited to a session during which they performed motor images of mental tasks one after the other, guided by visual cues. During the session, the participant is guided by clues in text form. The participant's task for each of these periods was to imagine for example in the second dataset: (1) raising the left hand (2) lowering the left hand; (3) Rise up right index finger; (4) Raise up right hand; (5) Pull down right hand; (6) Open the door; (7) Move a glass from point A to point B, (8) Close your eyes, (9) Think about red color, (10) Compute 11 × 12.

- Tools

The equipment that will be used for data acquisition is Emotiv EPOC. The acquired data may be different depending on the type of device used and the application for which it is intended. It is also important to know the characteristics of this tool. For the summary of the characteristics of the apparatus used (see Fig. 2a). For placement of the electrodes, the 10–20 system (Fig. 3) is used to describe the placement of the electrodes on a human scalp knowing that the brain divided into several subparts.

- Tasks

For this project, data collection sessions will be organized. The goal is to collect data from 10 volunteers. During the acquisition session, our volunteers will be put in normal conditions. They will be asked to imagine doing some of the actions listed below, but not physically doing them. Some of the actions requested of volunteers require meditation, and others are motor.

- The place

The brain is a complex organ and tasks such as vision, motor movements and emotions are treated in different places and so to avoid certain artefacts caused by certain movements, the subject will be asked to sit comfortably and relax normally. without also speaking. They will sit on a chair and in a room with less noise to avoid influencing him. Therefore, it was important to keep the place as quiet as possible.

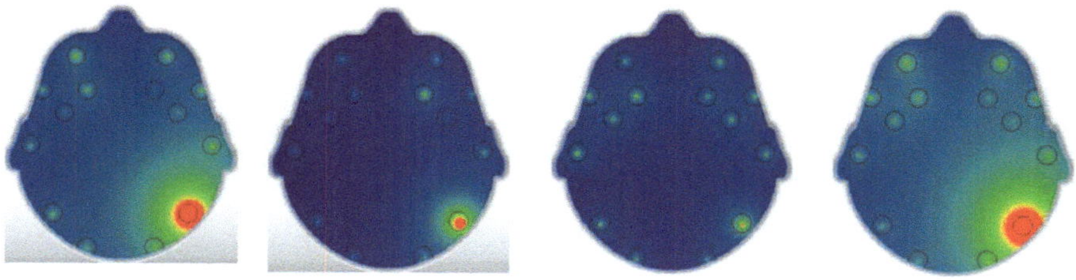

Fig. 11 Example of "Brainmap" EEG signal acquisition with 4 differents rythmes

- Structure of the dataset

The dataset contains 10 different actions. Each action here is considered as a class. During this session, 100 experiments per action have been recorded with 4 rhythms such as Delta, Theta, Alpha, Beta as shown in Fig. 11.

5.2 *Feature Extraction Method*

The feature extraction method used here consist on detecting first the center of each electrode by using Hough circle transform. Once the center coordinates C(x,y) are obtained, the radius R can be changed and then extract features from three geodesic disks defined by the center of the electrode C and a radius R. The features extracted in the geodesic disc are the mean, standard deviation and variance of intensity for data coming from each electrode.

5.3 *Results of Experiment*

In order to analyse the results of experiments (Table 5), the data are projected in an LDA space (see Fig. 12).

Table 5 Accuracy with SVM and Random forest

Model	Radius 7 (%)	Radius 14 (%)	Radius 21 (%)	All combined (%)
Random forest	81	89	92	91.1
SVM	74	81	86	87.3

The feature is extracted with different radius. The accuracies are given based on the radius

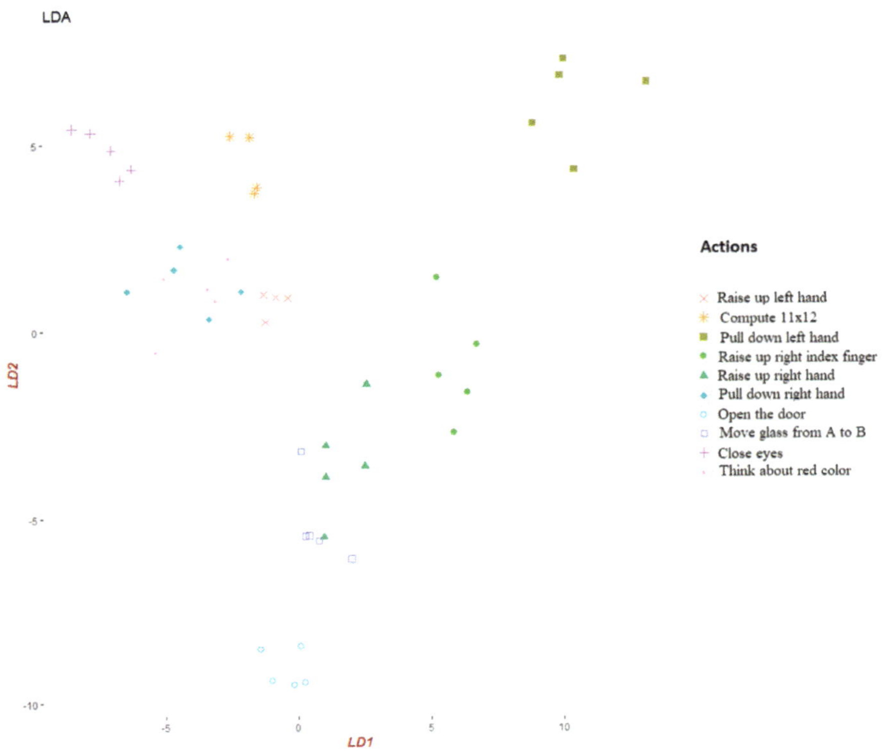

Fig. 12 LDA projection of the actions (intents) from one subject. *In this example, there are 10 differents actions represented with a shape and a color*

6 Conclusion

This chapter address a state of the art in EEG-based by reviewing the various techniques used from the protocols of data acquisition to classification to build a biometrics system for person's verification. In the state of the art, we have seen that different approaches have been used to address the authentication using EEG signals. The approaches can be different on the choice of number of electrodes used, on the pre-processing, on the feature extraction method and/or on the model used for the classification. EEG provides an answer to many applications.

We can also observe that the work that has been conducted to date gives very high accuracy. However, the observation that we must take into consideration is the lack of enough dataset available on this modality. It could allow a real comparison of the different approaches on the authentication based on the EEG.

References

1. Berger, H.: Über das elektrenkephalogramm des menschen. Arch. Psychiatr Nervenkr **87**, 527–570 (1929)
2. Berger, H.: Über das Elektrenkephalogramm des Menschen. XIII [The human electroencephalogram]. Archiv für Psychiatrie und Nervenkrankheiten **106**, 576–584 (1937)
3. Szurhaj, W., Lamblin, M.-D., Kaminska, A., Sediri, H.: EEG guidelines in the diagnosis of brain death. Neurophysiol. Clin./Clin. Neurophysiol. **45**(1), pp. 97–104, ISSN 0987-7053 (2015)
4. Jayarathne, I., Cohen, M., Amarakeerthi, S.: Survey of EEG-based biometric authentication. In: 2017 IEEE 8th International Conference on Awareness Science and Technology (iCAST), Taichung, pp. 324–329 (2017)
5. Basar, E., Düzgün, A.: How is the brain working?: research on brain oscillations and connectivities in a new "take-off" state. Int. J. Psychophysiol. **103**, 3–11 (2016)
6. Bassett, D.S., Gazzaniga, M.S.: Understanding complexity in the human brain. Trends in Cogn. Sci. **15**(5), 200–209, ISSN 1364-6613 (2011)
7. Sebastian023 [CC BY-SA 3.0 (https://creativecommons.org/licenses/by-sa/3.0)]
8. Sörnmo, L., Laguna, P.: Chapter 2—the electroencephalogram—a brief background. In: Sörnmo, L., Laguna, P. (eds.) In Biomedical Engineering, Bioelectrical Signal Processing in Cardiac and Neurological Applications, pp. 25–53. Academic Press, ISBN 9780124375529 (2005)
9. Kerbaj, D., Hassan, W., Nait-Ali, A.: Verifying a personas identity using brain responses to visual stimuli. In: 2017 2nd International Conference on Bio-engineering for Smart Technologies (BioSMART), Paris, pp. 1–6 (2017)
10. Marks, W.J.,, Laxer, K.D.: Chapter 7—invasive clinical neurophysiology in epilepsy and movement disorders. In: Aminoff, M.J. (eds.) Aminoff"s electrodiagnosis in clinical neurology (6th edn), pp. 165–185. W.B. Saunders. ISBN 9781455703081 (2012)
11. https://www.emotiv.com/, (online)
12. Kang, J.-H., Jo, Y.C., Kim, S.P.: Electroencephalographic feature evaluation for improving personal authentication performance. Neurocomputing, **287**, pp. 93–101, ISSN 0925-2312 (2018)
13. Chauhan, S., Arora, A.S., Kaul, A.: A survey of emerging biometric modalities. Procedia Comput. Sci. **2**, 213–218, ISSN 1877-0509. https://doi.org/10.1016/j.procs.2010.11.027 (2010)
14. Reshmi, K.C., Muhammed, P.I., Priya, V.V., Akhila, V.A.: A novel approach to brain biometric user recognition. Procedia Technol. **25**, 240–247, ISSN 2212-0173 (2016)
15. Eva, O.D., Lazar, A.M.: Comparison of classifiers and statistical analysis for EEG signals used in brain computer interface motor task paradigm. Int. J. Adv. Res. Artif. Intell. (IJARAI), **4**(1) (2015)
16. Voznenko, T.I., Chepin, E.V., Urvanov, G.A.: The control system based on extended BCI for a robotic wheelchair. Procedia Comput. Sci **123**, 522–527, ISSN 1877-0509 (2018)
17. Zhang, X., Yao, L., Sheng, Q., Kanhere, S., Gu, T., Zhang, D.: Converting Your Thoughts to Texts: Enabling Brain Typing via Deep Feature Learning of EEG Signals. https://doi.org/10.1109/percom.2018.8444575 (2018)
18. Nait-Ali, A.: Hidden biometrics: towards using biosignals and biomedical images for security applications. In: International Workshop on Systems, Signal Processing and their Applications, WOSSPA, Tipaza, pp. 352–356 (2011)
19. Rodrigues, R.N., Ling, L.L., Govindaraju, V.: Robustness of multimodal biometric methods against spoof attacks. J. Vis. Lang. Comput. **20**, 169–179 (2009)
20. Rao, T.K., Lakshmi, M.R., Prasad, T.: An exploration on brain computer interface and its recent trends. Int. J. Adv. Res. Artif. Intell. **1**, 17 (2012) https://doi.org/10.14569/ijarai.2012.010804
21. Sweeney, K.T., Ward, T.E., McLoone, S.F.: Artifact removal in physiological signals—practices and possibilities. IEEE Trans. Inf Technol. Biomed. **16**(3), 488–500 (2012)
22. Kumar, P.S., Arumuganathan, R., Sivakumar, K., Vimal, C.: Removal of artifacts from EEG signals using adaptive filter through wavelet transform. In: 2008 9th International Conference on Signal Processing, Beijing, pp. 2138–2141 (2008)

23. Kumar, P.S., Arumuganathan, R., Vimal, C.: Wavelet based ocular artifact removal from EEG signals using ARMA method and adaptive filtering. In: Proceedings—2009 IEEE International Conference on Intelligent Computing and Intelligent Systems, ICIS 2009, vol. 3 (2009). https://doi.org/10.1109/icicisys.2009.5358090
24. Croft, R.J., Barry, R.J.: Removal of ocular artifacts from the EEG: a review. J. Clin. Neurophysiol. **30**, 5–19 (2000)
25. Daly, I., Billinger, M., Scherer, R., Mueller-Putz, G.: On the automated removal of artifacts related to head movement from the EEG. IEEE Trans. Neural Syst. Rehabil. Eng. **21**(3), 427–434 (2013)
26. Ngoc, P.P., Hai, V.D., Bach, N.C., Van Binh, P.: EEG signal analysis and artifact removal by wavelet transform. In: Toi, V., Lien Phuong, T. (eds.) 5th International Conference on Biomedical Engineering in Vietnam. IFMBE Proceedings, vol. 46. Springer, Cham (2015)
27. Edla, D.R., Ansari, M.F., Chaundhary, N., Dodila, S.: Classification of facial expressions from EEG signals using wavelet packet transform and SVM for wheelchair control operations. Procedia Comput. Sci. **132**, 1467–1476, ISSN 1877-0509 (2018)
28. Puce, A., Hämäläinen, M.S.: A review of issues related to data acquisition and analysis in EEG/MEG studies. Brain Sci. **7**(6), 58 (2017)
29. Voznenko, T.I., Dyumin, A.A., Aksenova, E.V., Gridnev, A.A., Delov, V.A.: The experimental study of 'Unwanted Music' noise pollution influence on command recognition by brain-computer interface. Procedia Comput. Sci. **123**, 528–533, ISSN 1877-0509 (2018)
30. Zhou, S., Allison, B.Z., Kübler, A., Cichocki, A., Wang, X., Jin, J.: Effects of background music on objective and subjective performance measures in an auditory BCI. Front. Comput. Neurosci. **10**, 105 (2016). https://doi.org/10.3389/fncom.2016.00105
31. Motamedi-Fakhr, Shayan, Moshrefi-Torbati, M., Hill, Martyn, Hill, Catherine, White, Paul: Signal processing techniques applied to human sleep EEG signals—a review. Biomed. Signal Process. Control **10**, 21–33 (2014). https://doi.org/10.1016/j.bspc.2013.12.003
32. Chuang, J., Nguyen, H., Wang, C., Johnson, B.: I think, therefore I am: usability and security of authentication using brainwaves. In: International Conference on Financial Cryptography and Data Security, pp. 1–16. Springer (2013)
33. Jayarathne, I., Cohen, M., Amarakeerthi, S.: BrainID: development of an EEG-based biometric authentication system. In: 2016 IEEE 7th Annual Information Technology, Electronics and Mobile Communication Conference (IEMCON), Vancouver, BC, pp. 1–6 (2016)
34. Ngoc P.P., Hai, V.D., Bach N.C., Van Binh, P.: EEG signal analysis and artifact removal by wavelet transform. In: Toi, V., Lien Phuong, T. (eds) 5th International Conference on Biomedical Engineering in Vietnam. IFMBE Proceedings, vol. 46. Springer, Cham (2015)
35. Hazarika, N., Chen, J.Z., Tsoi, A.C., Sergejew, A.: Classification of EEG signals using the wavelet transform. In: Proceedings of 13th International Conference on Digital Signal Processing, Santorini, Greece, vol. 1, pp. 89–92. https://doi.org/10.1109/icdsp.1997.627975 (1997)
36. Altahat, S., Chetty, G., Tran, D., Ma, W.: Analysing the robust EEG channel set for person authentication. In: Arik, S., Huang, T., Lai, W., Liu, Q. (eds.) Neural Information Processing. ICONIP 2015. Lecture Notes in Computer Science, vol. 9492. Springer, Cham (2015)
37. Jayarathne, I., Cohen, M. Amarakeerthi, S.: BrainID: development of an EEG-based biometric authentication system. In: 2016 IEEE 7th Annual Information Technology, Electronics and Mobile Communication Conference (IEMCON), Vancouver, BC, pp. 1–6 (2016)
38. Poulos, M., Rangoussi, M., Alexandris, N.: Neural network based person identification using EEG features. In: 1999 IEEE International Conference on Acoustics, Speech, and Signal Processing. Proceedings. ICASSP99 (Cat. No. 99CH36258), Phoenix, AZ, USA, vol. 2, pp. 1117–1120 (1999)
39. Laka, P., Mazurczyk, W.: User perspective and security of a new mobile authentication method. Telecommun. Syst. (2018) https://doi.org/10.1007/s11235-018-0437-1
40. Blondet, M.V.R., Khalifian, N., Kurtz, K.J., Laszlo, S., Jin, Z.: Brainwaves as authentication method: proving feasibility under two different approaches. In: 2014 40th Annual Northeast Bioengineering Conference (NEBEC), Boston, MA, pp. 1–2. https://doi.org/10.1109/nebec.2014.6972734 (2014)

41. Pham, T., Ma, W., Tran, D., Nguyen, P., Phung, D.: A study on the feasibility of using EEG signals for authentication purpose. In: Lee, M., Hirose, A., Hou, Z.G., Kil, R.M. (eds.) Neural Information Processing. ICONIP 2013. Lecture Notes in Computer Science, vol. 8227. Springer, Berlin, Heidelberg (2013)
42. Jagadiswary, D., Saraswady, D.: Biometric authentication using fused multimodal biometric. Procedia Comput. Sci. **85**, 109–116, ISSN 1877-0509 (2016)
43. Abo-Zahhad, M., Ahmed, S.M., Abbas, S.N.: A new multi-level approach to EEG based human authentication using eye blinking. Pattern Recogn. Lett. **82**(2), pp. 216–225, ISSN 0167-8655 (2016)
44. Wang, M., Abbass, H.A., Hu, J.: Continuous authentication using EEG and face images for trusted autonomous systems. In: 2016 14th Annual Conference on Privacy, Security and Trust (PST), Auckland, 368–375 (2016)
45. Palaniappan, R.: Electroencephalogram signals from imagined activities: a novel biometric identifier for a small population. In: International Conference on Intelligent Data Engineering and Automated Learning, pp. 604–611. Springer (2006)
46. Zhang, X., Yao, L., Chen, K., Wang, X., Sheng, Q.Z., Gu, T.: DeepKey: an EEG and gait based dual-authentication system. CoRR abs/1706.01606 (2017): n. pag
47. Ashby, C., Bhatia, A., Tenore, F., Vogelstein, J.: Low-cost electroencephalogram (EEG) based authentication. In: 2011 5th International IEEE/EMBS Conference on Neural Engineering, Cancun, pp. 442–445 (2011)
48. Tektaş, F., Yücer, Ş. Kanak, A.: A new approach in border security applications with EEG biometrics. In: 2017 25th Signal Processing and Communications Applications Conference (SIU), Antalya, pp. 1–4 (2017)
49. Nakamura, T., Goverdovsky, V., Mandic, D.P.: In-ear EEG biometrics for feasible and readily collectable real-world person authentication. IEEE Trans. Inf. Forensics Secur. **13**(3), 648–661 (2018)
50. Singh, B., Mishra, S., Tiwary, U.S.: EEG based biometric identification with reduced number of channels. In: 2015 17th International Conference on Advanced Communication Technology (ICACT), Seoul, pp. 687–691 (2015)

Single Channel Surface EMG Based Biometrics

Samer Chantaf, Lobna Makni and Amine Nait-ali

Abstract An emerging biometric method based on surface EMG (SEMG) signals is considered. For this purpose, this chapter consists of two main parts. The first part reviews the SEMG signals in response to a force of fixed intensity from which frequencial parameters are extracted from the Power Spectral Density (PSD). The second part considers the M-wave signals muscle response following an electrical stimulation. M-wave signals are then characterized by extracting parameters using wavelet networks. The radial basis neural network (RBF) method is then used to classify these parameters.

1 Introduction

Generally speaking, Biometrics represent a secure alternative to traditional methods of individuals identity verification such as passwords. Biometrics use anatomical, physiological or behavioral features to determine the unique identity of individuals. Within this context, biometrics community deals commonly with some classical modalities such as fingerprint, palmprint, face recognition, iris, etc. However, due to the development of the technology of falsification, these modalities are subject to a potential spoofing. Therefore, the use of physiological signals such as the EMG is not very widespread in biometrics, considering the fact that they are very specific, distinctive and commonly used in the medical field. IN fact, Biometrics Research Group of the Lab. LiSSi at the University of Paris-Est (France) was one of the first groups who worked on this modality since 2009.

This chapter will be divided into two parts. In the first part, we present a method of biometric identification/verification based on the EMG signals of surface (SEMG) in response to a force of fixed intensity. In particular, we will first describe the sEMG, then we will describe the different processing steps that may be used to verify the

S. Chantaf
Lebanese University (LU), Tripoli, Lebanon

L. Makni · A. Nait-ali (✉)
Université Paris-Est, LISSI, UPEC, 94400 Vitry sur Seine, France
e-mail: naitali@u-pec.fr

A. Nait-ali (ed.), *Hidden Biometrics*, Series in BioEngineering,
https://doi.org/10.1007/978-981-13-0956-4_4

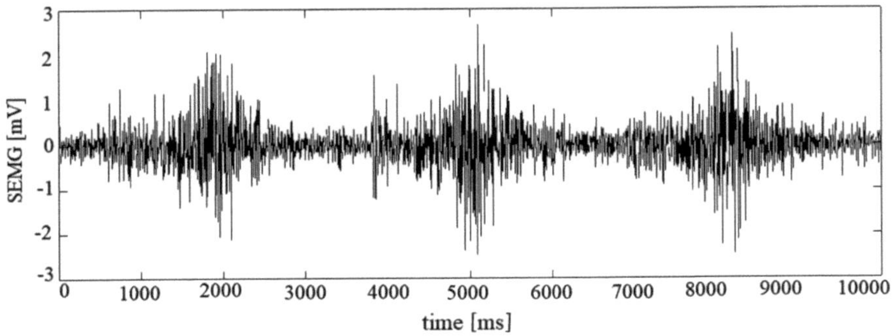

Fig. 1 Surface EMG signal

identity of individuals. In the second part of this chapter, we present an approach based on motor responses M, resulting from the study of motor nerve conduction.

2 Electromyogram

Generally speaking, EMG signals (Fig. 1) are weak biosignals (microvolts) that can be captured from contracted muscles. The muscles can be either neurologically or electrically activated. The procedure of measuring this electrical activity from the skin is called surface electromyography (sEMG) [1]. sEMG analysis provides various information about the peripheral nerves and muscles. For instance, it can be used to monitor the progress of some neuromuscular disorders. In other applications, the sEMG can be used to control mechanical systems such as exoskeleton. The low electrical signals generated by the active muscles, are detected by sensors placed on the skin, at the muscle level.

3 State of the Art: EMG as a Biometric Modality

In the field of biometrics, several research works based on EMG signals have been proposed using numerous processing approaches have been used. For instance, Chantaf et al. [2–5] recorded EMG signals from BIOPAC system. Afterwards, seven frequency domain features (e.g. average frequency, kurtosis, median frequency) are extracted and classified using RBF network. The system accuracy in [6] was 80%. In addition, Krishnamohan et al., used EMG signals measured at two channels (flexor carpi ulnaris and wrist) [7]. Then, they employed the non-uniform filter bank in order to extract frequency domain features which are reduced using vector quantization modeling. The authors obtained a recognition rate of their biometric system of about 83.94% [7]. Unlike Krishnamohan et al. [7] that used EMG signals from two channels, Kim et al. recorded signals from four channels (femoral rectus femoris, vastus medialis, vastus lateralis, and semitendinosus muscles) [8]. In contrast to [7], they extracted 12 time domain features (RMS, MAV, VAR, WAMP, ZC, SSC, IEMG, MMAV1, MMAV2, MAVSLP, SSI, and WL). In the classification step, all time domain features are classified using Artificial Neural Network. The identification rate of the system was 99.7%. Recently, the authors in [9] generated a personal feature vector that combines time (WL) and frequency (non uniform filter bank) domain features. In contrast to Krishnamohan et al. [7] that used vector quantization modeling for feature vector reduction, Kim and Pan employed PCA which reduced the feature vector dimension by projecting data using the principal axis of all components [9]. For classifier, they applied the LDA classifier in order to find the optimal projection in the feature space. Their authentication system showed the performance of 85%. Table 1 summarizes research works on biometric system.

In the same context, Yamaba et al. proposed a method that uses a list of gestures as a pass-gesture (i.e., password) [11, 12]. In fact, they demonstrated that the patterns of the same gestures of the same person are similar but they differ from those of other persons [12]. To identify pass-gestures, Yamaba et al. extract four time domain features which are maximum and minimum value of raw s-EMG and their associated time t_min and t_max [10]. For classification, SVM classifier of each subject were trained under these four features and tenfold cross validation was carried out using the same raw data. Unfortunately, experimental results in terms of FAR and FRR are not encouraging. Instead extracting features from EMG raw signal and use machine learning techniques, Kurogi et al. propose another method that reposes on DTW (Dynamic Time Warping) to compare pass-gestures and identify the specific pass-gesture of the specific person [13].

4 1st Case Study: Biometrics Using Surface EMG

In this approach, the Power Spectral Density (PSD) is calculated from the EMG responses. The PSD is then modeled and the extracted features are used as biometric information through a classifier (Neural Network RBF), as shown in Fig. 2.

4.1 Enrollment/Training Phase

- **Acquisition process**

The SEMG signals used are recorded from a BIOPAC system in response to a force of intensity of 40% of Maximum Voluntary Contraction (MVC). 10 healthy persons, mainly 9 men and 1 woman whose ages ranged between 25 and 40 years served as volunteers in this study. Our database includes 60 responses. The recorded responses

Table 1 Representation of research works on biometric system using EMG as behavioral feature

Research works	Feature extraction method	Reduced feature	Classifier	Recognition rate
Chantaf [6]	Seven frequency domain features (average frequency, kurtosis, median frequency, decile, coefficient dissymmetry, peak frequency)	–	RBFNN (Radial Basis Function Network)	80%
Krishnamohan et al. [7]	The non uniform filter bank method (frequency domain features)	vector quantization modeling	Gaussian Mixture Modeling (GMM)	83.94%
Kim et al. [8]	12 time domain features (RMS, MAV, VAR, WAMP, ZC, SSC, IEMG, MMAV1, MMAV2, MAVSLP, SSI, and WL).	–	Artificial Neural Network (ANN)	99.7%
Kim and Pan [9]	Non-uniform filter bank (frequency domain feature) and Waveform Length (time domain feature)	PCA and Linear Discriminant Analysis (LDA)	Euclidean Distance (ED), SVM, and KNN	85%
Yamaba et al. [10]	Four time domain features (max and min value of raw s-EMG and their associated time t_min and t_max)	–	SVM	–

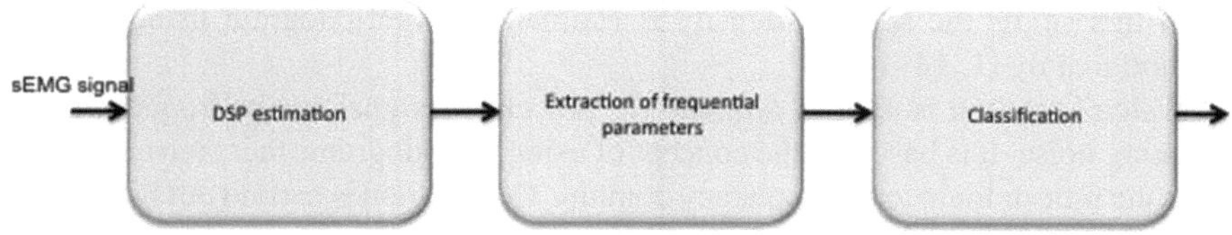

Fig. 2 General block diagram of the biometric approach by SEMG

consiste of 3 s of acquisition containing one contraction (SEMG signals) in respcnse to a clench force, and sampled at 2000 Hz.

The measurements are achieved as follows: three electrodes are placed on the right hand as shown in Fig. 3. The subject exercises a clench force of intensity 40% of MVC (Fig. 5) when a SEMG signal in response to this force of duration 3 s is recorded.

- **Feature extraction process**

The estimate of the Power Spectral Density (PSD) of an EMG signal is very important because it is calculated through the relevant parameters that should be used for the identification of individuals. A reminder of PSD is given in the following sub-section.

The power spectral density (PSD) describes how the power of a signal or time series is distributed with frequency. The goal of spectral density estimation is to estimate the spectral density of the signal from a sequence of time samples of the same signal. Conventionally, two estimation techniques exist: parametric and non-parametric. In our method, we used the non-parametric approach. Nonparametric methods are those in which the estimated PSD is made directly from the signal itself. The simplest method is the periodogram. It consists of calculating the discrete-time Fourier transform of sampled processes and then taking the magnitude squared of the result [14, 15].

Fig. 3 Measurements of SEMG showing clench force and electrodes position

In this study, the spectral density is estimated by periodogram using Welch's method mainly [1, 14–16].

Welch's method is used to estimate the power of a signal versus frequency, thus reducing noise. It is based on the concept of using periodograms that convert a signal from the time domain to the frequency domain. This method is carried out by dividing the time signal into successive blocks, then forming the periodogram for each block, and finally taking the average of all the blocks.

Each block is divided as follow (1):

$$x_i(n) = x(n + iD) \tag{1}$$

such that $n = 0, 1, \ldots M - 1$ and $i = 0, 1, \ldots L - 1$

Where M is the length of the blocks after division, D is the shifting between blocks and L is the number of blocks.

The periodogram for each block is given by (2):

$$\hat{S}^i(f) = \frac{1}{MU}\left|\sum_{n=0}^{M-1} x_i(n) \,.\, w(n) \,.\, e^{-j2\pi fn}\right|^2 \tag{2}$$

where U is the normalization factor of the window used to divide the signal into blocks (3).

$$U = \frac{1}{M}\sum_{n=0}^{M-1} w^2(n) \tag{3}$$

The Welch PSD estimate is given by (4):

$$\hat{S}_w(f) = \frac{1}{L}\sum_{i=0}^{L-1} \hat{S}^i(f) \tag{4}$$

Once the PSD is estimated, we extract the necessary parameters to be used in the classification and the verification of individuals. The extracted parameters are signal power, kurtosis, median frequency, deciles, dissymmetry coefficient and frequency peak.

- **Power of signal**

The power of a signal represents the distribution of energy M_0 (order 0) on the frequency axis (5).

$$M_r = 2\int_0^{\infty} f^r S_x(f)df \tag{5}$$

With S_x the estimation of the PSD by Welch method.

- **Average frequency**

Average frequency represents the statistical average of the signal.

$$MPF = \frac{M_1}{M_0} \tag{6}$$

- **Kurtosis**

Kurtosis measures the degree of peakedness of a distribution, defined as a normalized form of the fourth central moment μ_4 M_4 of a distribution (7).

$$CA = \frac{M_4^*}{M_2^{2*}} \tag{7}$$

- **Median Frequency**

The median divides the spectral density into two parts: 50% of data are less than the median, and 50% are greater.

The median is calculated by (8):

$$\int_0^{Fmed} S_x(f)df = \int_{Fmed}^{Fmax} S_x(f)df \tag{8}$$

- **Deciles**

We've seen that the median divides the distribution of the spectral density into two parts. We can generalize the division of this distribution into four, ten, one hundred, or n parts. The obtained values are named quartiles, deciles, percentiles or quantiles

$$\int_{f_F-1}^{f_F} S_x(f)df = k \int_0^{Fmax} S_x(f)df \quad 0 \preccurlyeq k \preceq 1 \tag{9}$$

Figure 4 shows how the deciles allow the testification of the fine form of the spectrum; the MPF represents its statistical average.

- **Coefficient of dissymmetry**

This parameter gives information about the shape of the spectral density from a symmetrical point of view. It is given by (10):

$$CD = \frac{M_3^*}{\sqrt{M_2^{3*}}}$$

Fig. 4 Parameterization of the spectrum

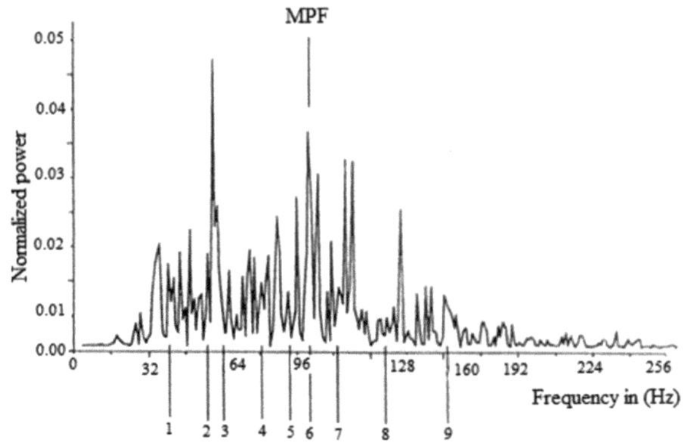

$$M_r^* = 2\int_0^\infty (f - MPF)^r S_x(f)df \tag{10}$$

- **Peak Frequency**

The peak frequency is the frequency for which the spectral density function is maximal. These parameters extracts are presented at the entrance of the artificial neural network RBF in order to identify the identity of individuals.

4.2 Test Phase

As described in the previous experiment (Section Acquisition of SEMG), the SEMG signals acquired resulted in 60 responses that consisted of 3 s of acquisition containing one contraction in response to a clench force and sampled at 2000 Hz. Once the PSD estimated, the necessary parameters are extracted and used in the classification and the verification of individuals. Figure 5a–d, show respectively the SEMG signal of 40% of intensity of MVC (Maximum Voluntary Contraction), estimated PSD in case of 40% of MVC (Maximum Voluntary Contraction) and the SEMG signal of 60% intensity of MVC (Maximum Voluntary Contraction), and estimated PSD in case of 60% of MVC (Maximum Voluntary Contraction).

Once extracted, these parameters are presented to the input of the artificial neural network to be classified. Radial Basis Function (RBF) neural network is used in supervised applications. They are embedded into a two-layer feed forward neural network.

This network is characterized by a set of inputs and a set of outputs where a hidden layer lies in between (Fig. 6). The neurons in the hidden layer contain Gaussian transfer functions (Fig. 7). To train RBF networks we have to determine the number of neurons in the hidden layer, the coordinates of the center of each hidden-layer, and the weights applied to the RBF function outputs [2–4, 14–18].

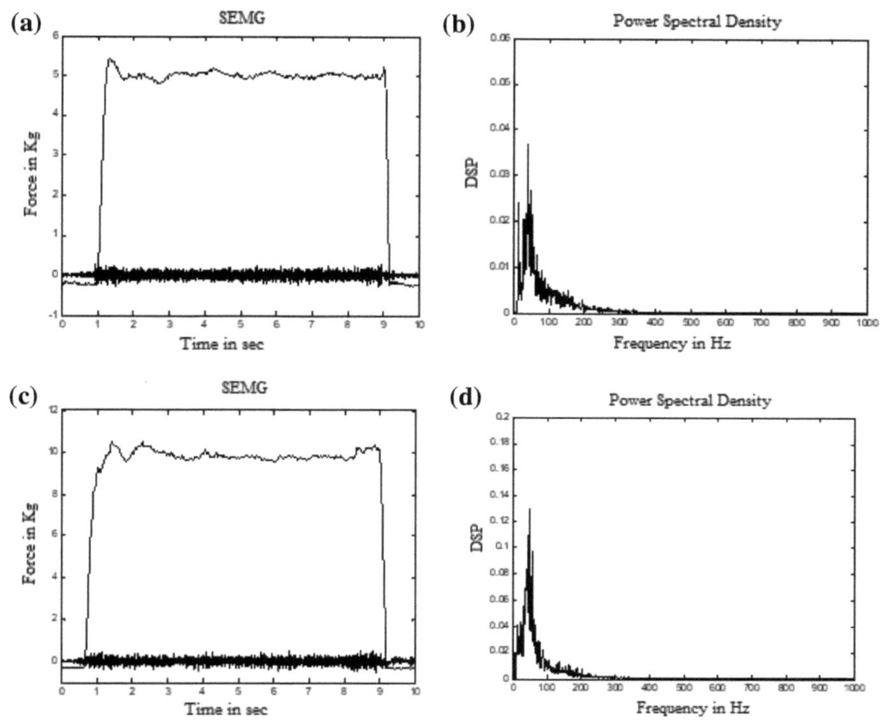

Fig. 5 **a** SEMG of 40% intensity of MVC **b** Estimated PSD of 40% of MVC **c** SEMG of 50% intensity of MVC **d** Estimated PSD of 60% of MVC

The training consists of optimizing the network parameters in order to fit the network outputs to the given inputs [4].

The output of the network is:

$$\varphi(x) = \sum_{i=1}^{N} a_i \rho(||x - c_i||) \tag{11}$$

The activation function used for the neurons in the hidden layer is the Gaussian function. It is given by (12):

$$\rho(||x - c_i||) = exp\left[-\beta ||x - c_i||^2\right] \tag{12}$$

where x is the input feature vector, N is the number of neurons in the hidden layer, c_i is the center vector for neuron i, and a_i are the weights of the linear output neuron. The weights a_i, c_i and β are determined in a manner that optimizes the fit between φ and the data.

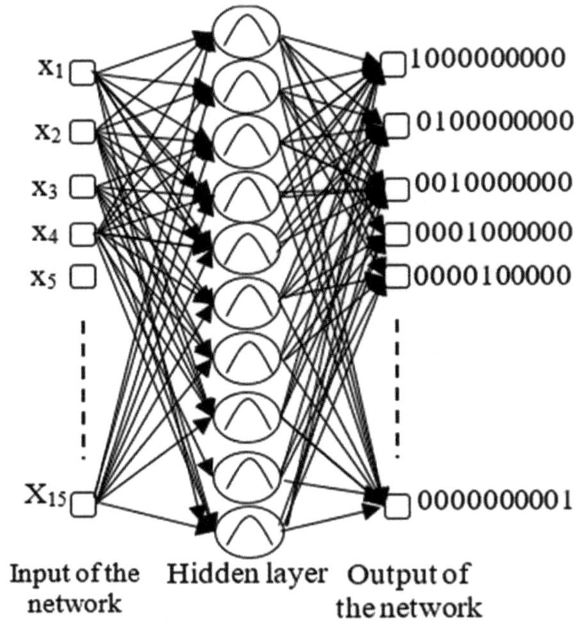

Fig. 6 Structure of the network RBFNN

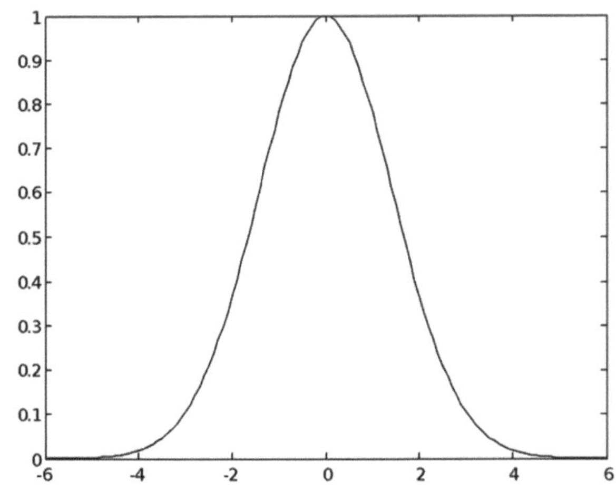

Fig. 7 Gaussian function

In this work, the neural network inputs correspond to the parameters extracted from the PSD (15 parameters) and the outputs correspond to each specific individual.

The training set is composed of 30 contractions for each individual and the neural network was learned for 200 epochs. Then the neural network is tested using other contractions from the same individuals.

The extracted parameters are presented to the input of the network to be classified.

Table 2 shows the classification of performances corresponding to each individual.

Table 2 Results of the identification performance in case of clench force of 40% intensity of MVC

Number of contractions	Person 1 (%)	Person 2 (%)	Person 3 (%)	Person 4(%)	Person 5 (%)
10	60	50	50	50	50
15	73	67	67	73	67
20	75	75	75	75	75
25	76	76	76	76	76
30	80	80	77	80	77
Number of contractions	Person 6 (%)	Person 7 (%)	Person 8 (%)	Person 9 (%)	Person 10 (%)
10	60	50	50	50	60
15	67	73	73	67	67
20	75	75	75	75	75
25	75	76	76	76	76
30	80	80	80	77	80

Each performance is analyzed by using contractions in response to a force of 40% intensity of MVC, and by modifying, each time, the number of contractions employed to identify individuals (10, 15, 20, 25, and 30).

Results from Table 2 show that the highest identification performance for each individual is obtained when using 30 contractions for the authentication process. The performance increases up to 80% when using 30 contractions [5]. But this performance remains constant (80%) when using more than 30 contractions. Consequently, the best results are achieved at 30 contractions.

Table 3 shows the classification of performances corresponding to each individual. The performance is analyzed by using contractions in response to a force of 60% intensity of MVC, and by modifying each time the number of contractions employed to identify individuals (10, 15, 20, 25, and 30).

Results from Table 3 show that the highest identification performance for each individual is obtained when using 30 contractions for the authentication process. The performance increases up to 70% when using 30 contractions [5]. But this performance remains constant (70%) when using more than 30 contractions. Consequently, the best results are achieved at 30 contractions.

Therefore, we note from Tables 1 and 2 that the performance obtained in response to a force of 40% intensity of MVC is better than the one obtained in response to a force of 60% intensity of MVC. This difference in performance is explained by the fact that at high currents, a phenomenon of fatigue is observed. Thus, the energy produced by the muscle fluctuates producing a change in the frequency content of the EMG signal [14].

Table 3 Results of the identification performance in case of clench force of 60% intensity of MVC

Number of contractions	Person 1 (%)	Person 2 (%)	Person 3 (%)	Person 4 (%)	Person 5 (%)
10	40	30	30	30	40
15	47	53	47	47	47
20	55	60	60	60	55
25	60	64	64	64	60
30	70	67	70	67	70
Number of contractions	Person 6 (%)	Person 7 (%)	Person 8 (%)	Person 9 (%)	Person 10 (%)
10	30	40	30	40	30
15	53	47	53	47	53
20	60	55	60	55	60
25	64	60	64	60	64
30	70	67	67	70	67

4.3 Performance Evaluation and Analysis

The technique of biometric identification proposed is analyzed in the two different cases: (1) Effect of fatigue (2) Effect of the movements of the electrodes on the performance of identification obtained in response to a force of 40% intensity of MVC.

- **Effect of fatigue**

In the first case, an individual are asked to execute many (push-ups).

Figure 8 shows that fatigue had no significant effect on the authentication perfor-

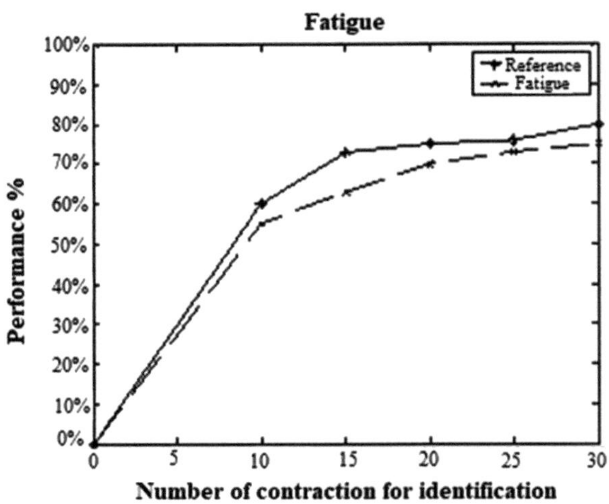

Fig. 8 Fatigue after push-ups in case of 40% of MVC

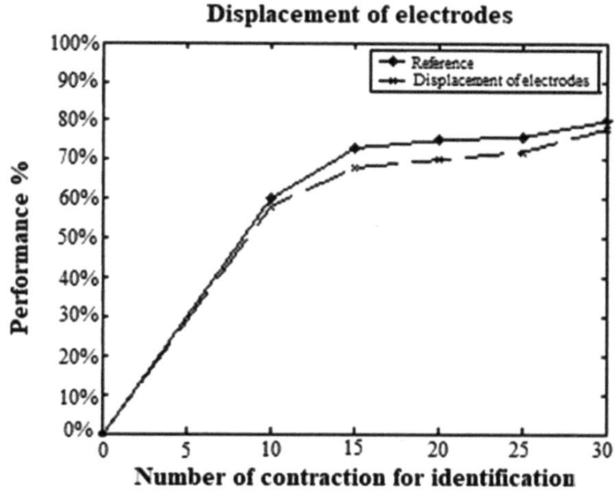

Fig. 9 Sensitivity of performance compared to the displacement of the electrodes

mance since the curves were very close.

- **Effect of the displacement of electrodes**

In the second case, the electrodes were moved 3 cm from their position. As shown in Fig. 9, we notice that such displacement had no effect on the performance.

5 2nd Case Study: EMG Biometrics Using Motor Response M

5.1 Enrollment/Training Phase

Generally speaking, to examine the conduction properties of motor nerve fibers, a mixed nerve is electrically stimulated with a single, short duration stimulus. The action potentials are generated in alpha motoneurons axons which innervate extrafusal muscle fibers of skeletal muscle and are directly responsible for initiating their contraction.

The evoked nerve action potentials propagate along the motor fibers and activate the neuromuscular junctions of the stimulated motor axons. Synaptic transmission at the neuromuscular junctions between motor axons and muscle fibers produces muscle action potentials on innervated muscle fibers and triggers a twitch contraction in them. In motor NCS, the M wave that precedes the actual muscle contraction is measured.

The amplitude of this wave is proportional to the number of muscle fibers depolarized, and to the reflection of the extent of activation of muscle fibers which are produced by motor nerve stimulation.

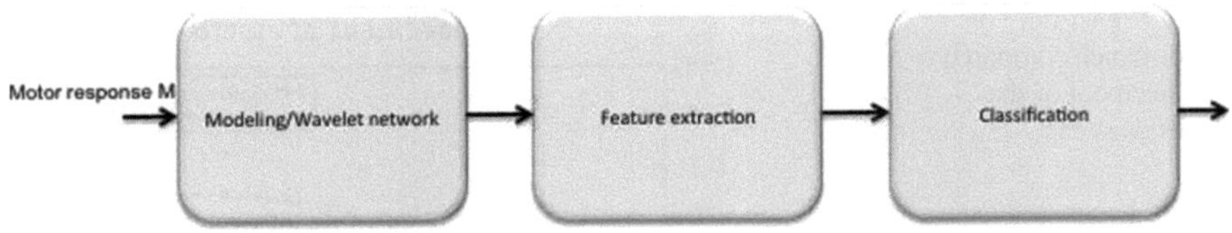

Fig. 10 Block diagram showing the main steps required to verify individuals by their M-wave response

As the current amplitude increases during motor nerve stimulation, more motor axons are progressively recruited and a maximum twitch contraction is produced. At this point, all the stimulated motor axons are excited, the maximum number of neuromuscular junctions is triggered, and action potentials are generated on the membranes of all innervated muscle fibers [19]. Any increase in stimulus amplitude will not increase the amplitude of the M response.

In this part, the M wave is presented as biometric tool used to determine the identity of individuals. This method consists of modeling the M-waves by wavelet network in order to extract the significant parameters (translation, dilation and weight) from different wavelets.

The different phases are described in the following sub-section (Fig. 10).

5.2 *Acquisition Process*

The M-waves are recorded from 10 healthy individuals (8 males and 2 females) by the KEYPOINT system at the Rafic Hariri public hospital. The M-waves are obtained in response to an electrical stimulation of 20 and 30 mA intensities at different moments during nine months.

Our database is composed of ten sets for each person: five folders of M-waves in response to a stimulation of each intensity (20 and 30 mA). Each set contains 20 simultaneous responses sampled at 1 Hz.

The signals are collected by the intermediary of three electrodes placed on the right hand as shown in Fig. 11a. 20 electrical stimulations are successively performed on the motor nerve of the hand (Fig. 11b). The response M is recorded from the thénar (Fig. 11c).

5.3 *Feature Extraction Process*

The modeling of motor responses is the core of our method. The M waves responses are modeled by wavelets networks.

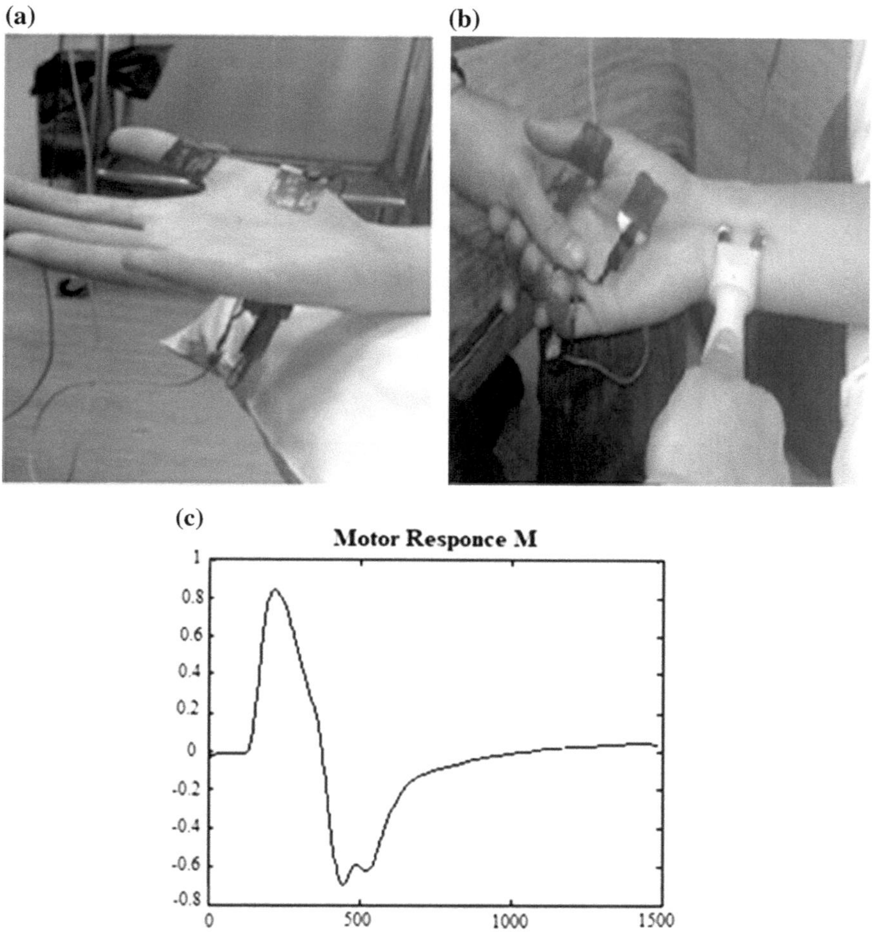

Fig. 11 **a** Position of electrodes; **b** Stimulation; **c** Motor Response M

The wavelets networks are a class of neural networks. They are the result of the combination of the theory of wavelets with multilayer neural networks of feedforward type. The wavelets networks utilize the functions of the wavelets as functions of activation.

The methods of construction of the wavelets network are developed using the theoretical characteristics of the transformed wavelets. Each response is modeled by 8 wavelets. The used wavelet is the inverse Mexican Hat (Fig. 12).

The parameters (translation, dilatation and weight) are extracted from different wavelets. These parameters are used to identify individuals (Fig. 13a, b).

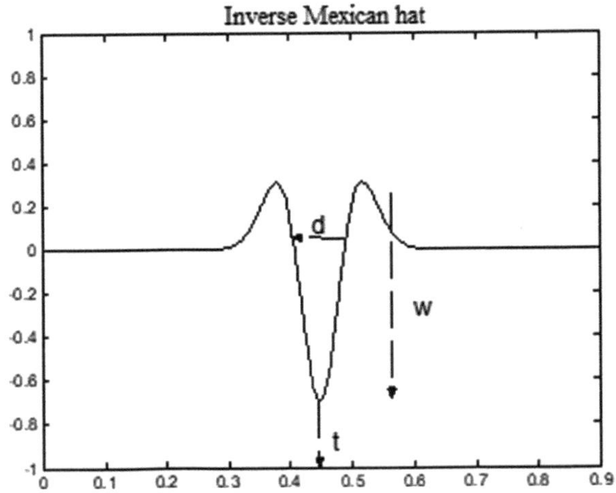

Fig. 12 Inverse Mexican hat

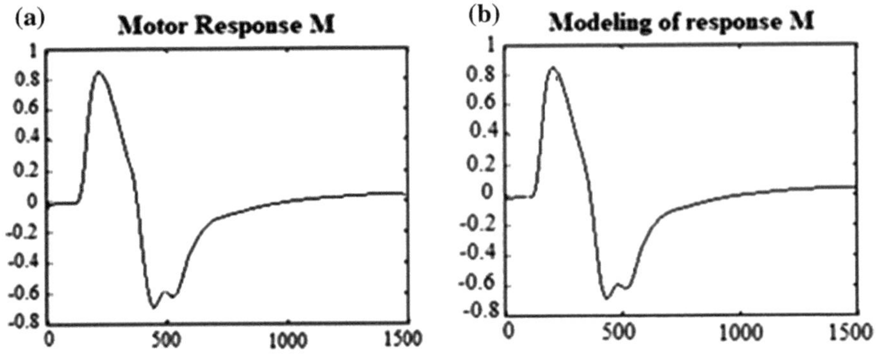

Fig. 13 **a** Motor response M; **b** Modeling of the response M

5.4 *Test Phase*

The classification is carried out by the network RBF. The input of the network is comprised of the extracted parameters of different wavelets. Our network is driven by 15 M wave for each individual. The identification is done by changing the number of responses used: 5, 10 and 15 responses.

The RBF network consists of a layer of input containing 24 entries corresponding to the parameters extracted from the wavelet, a hidden layer to 10 neurons and a layer of output. The number of outputs of the neural network is equal to the number of individuals to be identified (Fig. 14). The activation function used is the Gaussian function (Fig. 7).

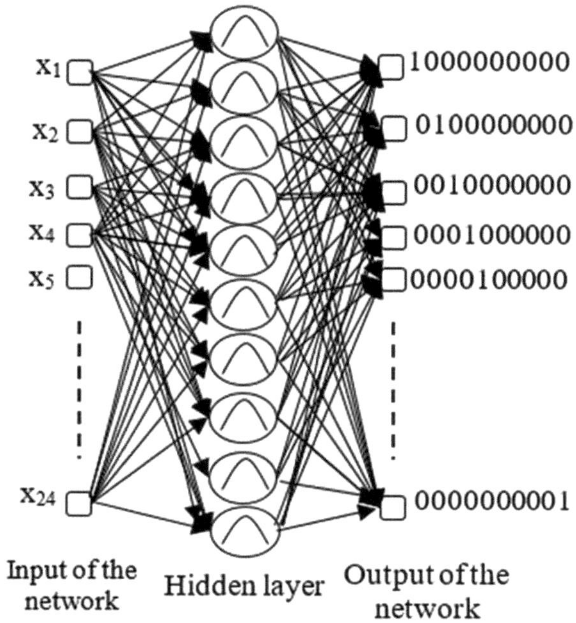

Fig. 14 Structure of the network RBF

5.5 *Performance Evaluation and Analysis*

Table 4 shows the classification performances corresponding to each individual in case of stimulation of 20 mA intensity. It shows that the highest identification performance is obtained when using 15 responses for the identification. The performance increased up to 100%. The performance is analyzed by modifying the number of responses to identify individuals (5, 10 and 15 responses) each time.

Table 4 Results of the identification performance in case of stimulation of 20 mA intensity

Number of responses	Person 1 (%)	Person 2 (%)	Person 3 (%)	Person 4 (%)	Person 5 (%)
5	80	100	80	80	80
10	90	100	90	90	90
15	93	100	93	93	93
Number of responses	Person 6 (%)	Person 7 (%)	Person 8 (%)	Person 9 (%)	Person 10 (%)
5	80	100	100	80	80
10	90	100	100	90	90
15	93	100	100	93	93

Table 5 Results of the identification performance in case of stimulation of 30 mA intensity

Number of responses	Person 1 (%)	Person 2 (%)	Person 3 (%)	Person 4 (%)	Person 5 (%)
5	80	100	100	80	80
10	90	100	100	90	90
15	93	100	100	93	93
Number of responses	Person 6 (%)	Person 7 (%)	Person 8 (%)	Person 9 (%)	Person 10 (%)
5	80	100	100	80	80
10	90	100	100	90	90
15	93	100	100	93	93

Table 5, corresponding to the case of stimulation of 30 mA intensity, shows also that the highest identification performance is 100% and it is obtained when using 15 responses during the identification process. Consequently, the best performance is obtained when using 15 responses for the identification regardless of the intensity of the stimulation [5].

Therefore, according to the results obtained from Tables 4 and 5, we conclude that the best performance is obtained when one uses 15 M wave in the identification phase. This performance varies slightly with the intensity of stimulation. In fact, it is found that the performance is slightly higher in the case of stimulation at 30 mA than the one of stimulation at 20 mA. This is explained by the fact that when we increase the amplitude of electrical stimulation, the number of motor axons stimulated increases and therefore, the amplitude of the wave M increases. Also, the amplitude of the wave M increases with the current of stimulation until a maximum contraction occurs. At this level of stimulation, the amplitude of the wave M does not grow.

6 Conclusion

In the field of biometrics, the shape of the hand, iris, the fingerprint or the details of the face are the most used techniques, among others. But once these characteristics change, security is no longer assured. The approach proposed in this thesis is highly specific and reliable since the recorded signals cannot be voluntarily modified (unlike a fingerprint for which a simple wound cancels its validity). In the course of this work, we have developed two methods of biometric identification based on the physiological signals EMG and motor response M.

The first highlighted method uses the EMG signals in response to a force with a fixed intensity of 40% of MVC and 60% of MVC to identify the different individuals. This technique consists of calculating the power spectral density of the EMG signals by the method of Welch and then extracting the frequential parameters, the most significant to the biometric identification such as power, the kurtosis, median frequency, the deciles, the coefficient of asymmetry and the frequency peak. The classification is carried out by the network RBF, Which is made up of an input layer containing 15 entries corresponding to the extracted parameters of the PSD, a hidden layer composed of 5 neurons including the activation function used (Gaussian function) and an output layer in which the number of outputs is equal to the number of individuals to identify. 30 contractions are used for learning the network. This technique is insensitive to noises and provides 80% performance.

The last method of biometric identification proposed uses the motor responses M. The responses are obtained as a result of electrical stimulation of 20 and 30 mA intensities. The modeling techniques and classification based on wavelets networks and neural networks (RBF) were implemented. The parameters of classification are extracted from the wavelets used in the modeling (namely translation, dilatation and weight). The network RBF used is composed of a input layer containing 24 entries corresponding to the parameters extracted from the wavelet, a hidden layer 5 neurons, function of Gaussian activation and an output layer in which the number of outputs is equal to the number of individuals to identify. 15 motor responses are used in the learning phase and 95% performance was obtained which is considered as an interesting recognition rate. In addition, the statistical studies show that our method is not sensitive to any type of noise, thus making this technique particularly vital.

References

1. Marple, S.L.: Digital Spectral Analysis with Applications. Prentice-Hall, Englewood Cliffs, NJ, USA (1987)
2. Chantaf, S., Naït-Ali, A., Karasinski, P., Khalil, M.: ECG modelling using wavelet networks: application to biometrics. Int. J. Biometrics **2**(3), 236–249 (2010)
3. Rumelhart, D.E., Hinton, G.E., Williams, R.J.: Learning representations by back propagation errors. Nature 323, 533–536 (1986)
4. Poggio, T., Girosi, F.: Networks for approximation and learning. Proc. IEEE **78**(9), 1481–1497 (1990)
5. Chantaf, S.: Biométrie par signaux physiologiques. Ph.D. Thesis. Université Paris Est Créteil (UPEC), Paris, France (2011)
6. Chantaf, S.: Thesis, Biométrie par signaux physiologiques (2011)
7. Krishnamohan, P., Holi, M.S., et al.: Gmm modeling of person information from emg signals. In Proceeding of the 2011 IEEE Recent Advances in Intelligent Computational Systems (RAICS'11), Trivandrum, India, pp. 712–717 (2011)
8. Kim, S.-H., Ryu, J.-H., Lee, B.-H., Kim, D.-H.: Human identification using emg signal based artificial neural network. J. Inst. Electron. Inf. Eng. **53**(4), 142–148 (2016)
9. Kim, J.S., Pan, S.B.: A study on EMG-based biometrics. J. Internet Serv. Inf. Secur. **7**(2), 19–31 (2017)
10. Yamaba, H., Kurogi, T., Aburada, A., et al.: On applying support vector machines to a user authentication method using surface electromyogram signals. Artif. Life Robot. **22**, 1–7 (2017)
11. Yamaba, H., Kurogi, T., Kubota, S., et al.: An attempt to use a gesture control armband for a user authentication system using surface electromyograms. Int. Symp. Artif. Life Rob. 342–245 (20160
12. Yamaba, H., Kurogi, T., Kubota, S., et al.: Evaluation of feature values of surface electromyograms for user authentication on mobile devices. Artif. Life Robot. **22**, 108–112 (2017)

13. Kurogi, T., Yamaba, H., Aburada, K., Katayama, T., Park, M., Okazaki, N.: A study on a user identification method using dynamic time warping to realize an authentication system by s-EMG. Int. Conf. Emerg. Intell. Data and Web Technol., 889–900 (2018)
14. Kay, S.: Modern Spectral Estimation Theory and Application. Englewood Cliffs, Prentice-Hall, NJ, USA (1988)
15. Marple, S.L.: Digital Spectral Analysis with Application. Englewood Cliffs, Prentice-Hall, NJ, USA (1987)
16. Proakis, J.G., Manolakis, D.G.: Digital Signal Processing Principles, Algorithms, and Applications. Englewood Cliffs, Prentice-Hall, NJ (1996)
17. Haykin, S.: Neural Networks: A Comprehensive Foundation, 2nd edn. Macmillan College Publishing Company, New York (1998)
18. Jain, A.K., Ross, A., Prabhakar, S.: An introduction to biometric recognition. IEEE Trans. Circuits Syst. Video Technol. **14**(1), 4–20 (2004)
19. Robinson, A.J., Snyder-Mackler, L.: Clinical electrophysiology: electrotherapy and testing procedures. In: 2007, Chapter 12: Clinical Electrophysiologic Examination and Evaluation, 3rd edn, p. 424, Lippincott Williams and Wilkins, Baltimore, MD
20. Yamaba, H., Nagatomo, S., Aburada, K., et al.: An authentication method for mobile devices that is independent of tap-operation on a touchscreen. J. Robot. Netw. Artif. Life **1**, 60–63 (2015)

Wearable Multi-channel EMG Biometrics: Concepts

Ikram Brahim, Islame Dhibou, Lobna Makni, Sherif Said and Amine Nait-ali

Abstract In this chapter, a case study using a specific wearable Multi-Channel EMG device will be considered. In particular, eight EMG channels will be used through Myo Armband system. The purpose is to deploy a verification biometric system using EMG signals corresponding to hand gestures. More specifically, the idea behind this concept is the capacity to generate a digital signature for each specific hand gesture.

1 Introduction

As we saw in the previous chapter, the purpose of the study was to explore a concept of EMG (electromyogram) biometrics as a potential modality that can be used to verify individuals using a single acquisition channel. As an extension to this basic concept, multi-channel EMG acquisitions systems can have many advantages in terms of accuracy and noise reduction. For example, in such systems, the Signal-to-Noise (SNR) ratio can be improved using numerous signal processing approaches such as: averaging, source separation, filtering and decomposition approaches, etc.

In this chapter, we will consider a case study using a specific wearable multi-channel EMG device, called Myo Armband. The purpose is to discuss a verification biometric system using EMG signals corresponding to hand gestures. In fact, the idea behind this concept is to generate a digital signature for each specific hand gesture. This signature can be used for both the enrollment phase and the verification phase.

This chapter is organized as follows: in Sect. 2, some generalities on Myo system are presented. In Sect. 3, an EMG individual verification protocol using Myo Armband device is described. This concept is then extended to a general purpose system dealing with hand gestures recognition. Finally, a conclusion is provided in the last Section.

I. Brahim · I. Dhibou · L. Makni · S. Said · A. Nait-ali (✉)
Université Paris-Est, LISSI, UPEC, 94400 Vitry sur Seine, France
e-mail: naitali@u-pec.fr

S. Said
American University of Middle-East (AUM), Egaila, Kuwait

A. Nait-ali (ed.), *Hidden Biometrics*, Series in BioEngineering,
https://doi.org/10.1007/978-981-13-0956-4_5

2 Myo System: A Multi-channel EMG System

EMG signal is an biosignal generated by muscle cells once they are electrically or neurologically activated. These signals can be acquired through different sensors. On the other hand, Myo device is a wearable bracelet that can be worn on the forearm, just below the elbow. It has been developed by Thalamic Labs in 2014 [1].

This bracelet can expand between 7.5 and 13 inches (19–34 cm) forearm circumference (see, Fig. 1). It weighs 93 grams which gives it the ability to be comfortably be worn for a long time.

The structure of Myo consists of eight EMG (electromyography) sensors and an Inertial Measurement Unit (IMU) which consists of a gyroscope, accelerometer and a magnetometer. More specifically, the Myo can detect electrical activation (contraction) produced by the muscle. It detects the electrical activity in forearm muscles below the elbow. In fact, the human forearm has different types of muscles. These muscles control the movement of the wrist, such as moving fingers, making a fist, turning left or right, etc.

Numerous applications have been developed using Myo device. Commonly, it is employed in control applications. For example, it can be used to achieve simple actions such as light control or door opening, but can be also employed in some advanced control applications such as drones control and robotics, including in the medical field.

By considering our context, Myo Armband device will be explored in biometrics application to recognize individuals or to recognize gestures. This is the purpose of the next two sections.

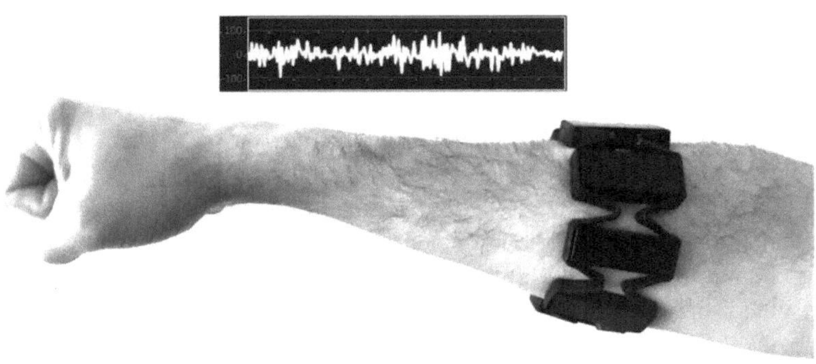

Fig. 1 A person wearing Myo device for biometrics application

3 EMG Biometrics Modality for Person Verification: Concept

3.1 EMG Verification Generic Scheme/Protocol

As in any verification system, the proposed system uses a generic scheme where two phases are required, namely: (1) enrollment phase, (2) verification phase (Fig. 2). The purpose of this application is generate a unique signature (as a password) for each individual from successive hand gestures. Technically, the EMG signals are recorded from Myo device, connected wirelessly to a computer in which a signal processing algorithm is implemented. In the protocol we considered, each user is asked to produce a number of successive hand gestures.

Each hand gesture will activate some muscles, and generates hence eight signals corresponding to eight sensors (see Fig. 3). From the experiments, it is obvious that similar hand gestures will not produce exactly the same signal patterns. This may due

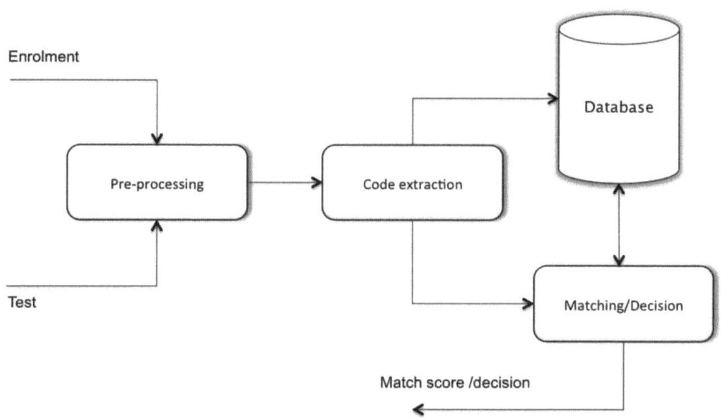

Fig. 2 Biometrics generic scheme

Fig. 3 Verification approach using Myo-armband

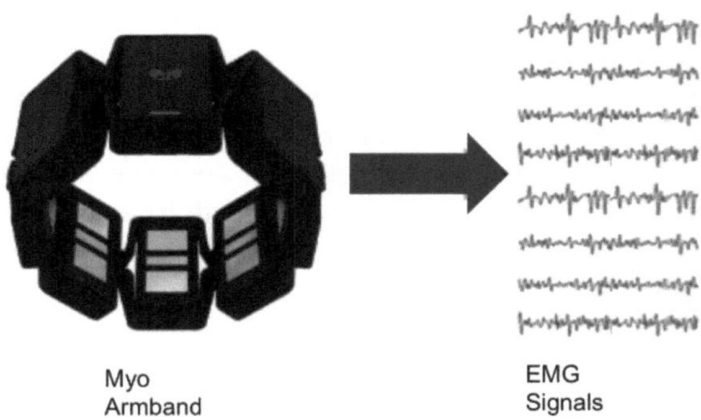

for many reasons. For example, the intensity for each muscular contraction will not be perfectly identical for the same person. The fatigue factor, or even the posture can impact the reproducibility of the patterns. However, by analysing the signals, one can notice that intra-individual similarities are observed in some channels. These similarities should be be identified before the enrollment phase to ensure a stability property. On the other hand, inter-individual disparities can also be observed when considering the same gestures which make the modality an interesting candidate to be used for person recognition. From a collection of experiments, on can determine a sort of dictionary of potential explotable gestures. In our case, we explored the following gestures:

- Fist
- Double tap
- Hand in
- Hand out
- Hand down
- Three

A password is then created by a combination of these gestures.

3.2 *Enrollment/Training Phase*

In this phase, the user memorizes and reproduces successively a set of gestures as if he was typing a password. For example, one can consider the following gesture combination: Hand in, Hand out and Three. Since, the Myo device captures eight EMG signals, each one is pre-processed (pre-processing phases) by a simple energy calculation over a specific observation (activation) window. If the energy is lower than a threshold, the signal is associated to a low level. However, if the signal energy is above a threshold, the signal level is associated to a high level. Figure 4 shows a hand-in gesture where the recorded signals are quantified (or converted) into two levels signals (High-Low) according to a specific threshold. Once normalized, one can notice that only energies above 70% are associated to a high signal level (Fig. 5).

Consequently, in the second step of the biometric scheme, the high level will be associated to the digit “1”, and the low level to “0”. Therefore, a single gesture will generate eight digits and the whole combination of gesture will generate 8xN digits, where N is the number gestures used by the user to generate a specific password. For instance, if the user generates a password from a combination of three hand gestures, the corresponding length password will be 24 binary digits (Fig. 6).

In this case, it is important to mention that the observation window duration for each gesture should be defined by the protocol itself. Users can either generate long or short transition movements according to a given protocol. During the enrolment process, the user can asked to produce many instance of the same combination of the hand gesture. For example, collecting ten trials of hand gestures showed that the

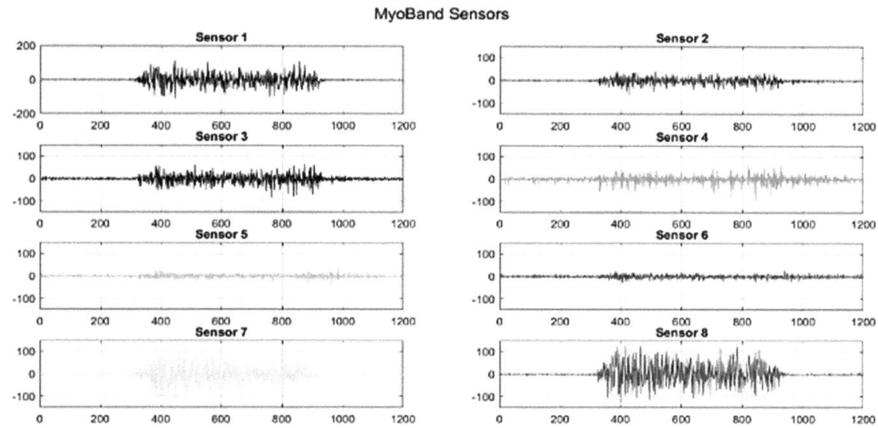

Fig. 4 Signals from eight sensors for hand-in gesture

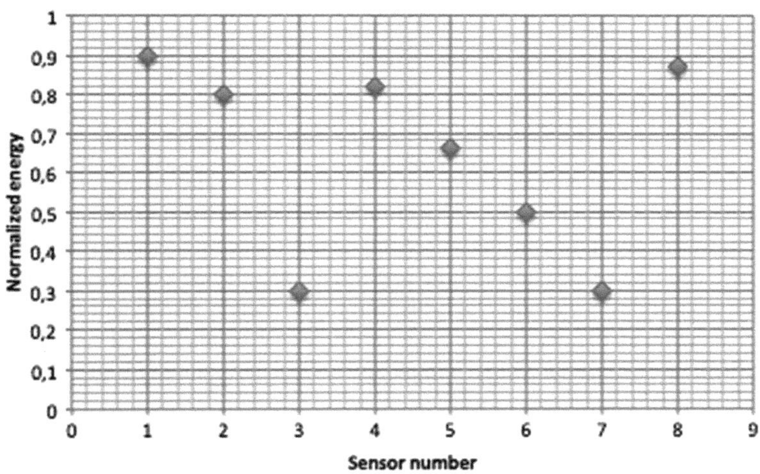

Fig. 5 Normalized signal for hand-in gesture and thresholding process. Only energies above 70% are associated to a high signal level

same gesture of the same person activates repeatedly the same muscles for which the EMG is captured by an electrode.

The collection of the binary codes are saved in a database, or the most dominant pattern is saved as a signature.

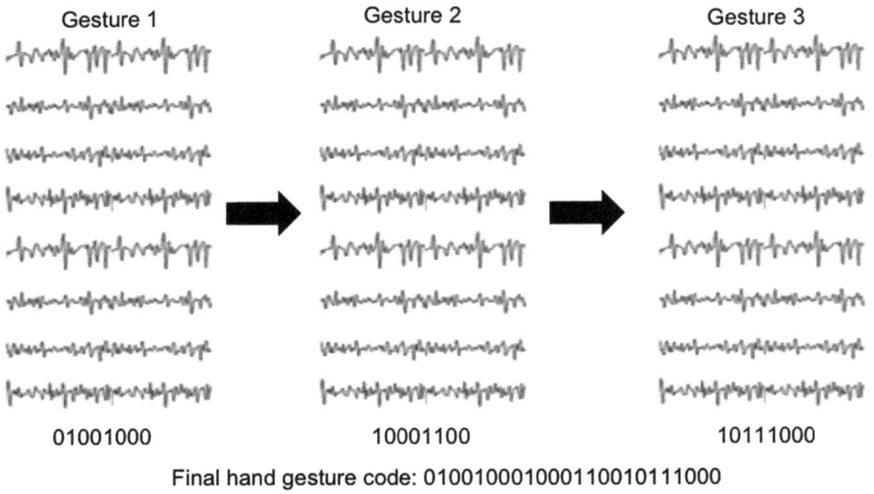

Fig. 6 Example of encoding a combination of three hand gestures producing a 24 digits binary code

3.3 Test Phase

During the test phase, the user who aims to access to a resource using the EMG based biometrics system should claim his identity (i.e. login) then reproduce a combination of hand gestures similar to the ones used during the enrollment phase (i.e. password). Naturally, each hand gesture should previously have been encoded in the initial dictionary. In addition, as specified, the duration for each movement should follow the same rhythm as the one used for the enrollment. Usually, a well trained user can be comfortable to reproduce approximately the same movements at the same rhythm (Speed) for each trial.

As in any verification system, the generated code is compared to the corresponding code saved in the database. A matching rule should be defined in order to grant or deny the access to the resources.

4 EMG for Gesture Recognition

In this section, the concept of using Myo device is restricted to a general purpose hand gesture recognition. It differs from the previous concept in the sense that the aim is to recognize the gesture regardless the identity of the user, as seen previously. Englehart et al. [2] argued that the recognition of multiple hand gestures could be considerably improved using the EMG multi-channel data rather than using a single-channel EMG data. In fact, gesture recognition based on EMG signals is a challenging task due to its broad applications in myoelectric control. It has been treated as a pattern recognition

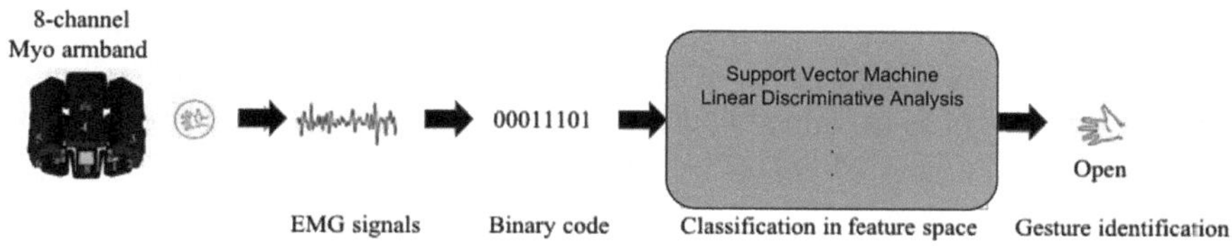

Fig. 7 Gesture recognition system showing the path from a hand gesture until the classification phase

problem in several research works [3–5]. Generally speaking, human motions can be captured by EMG signals which can be intelligently recognized as control commands in many myoelectric systems like prosthesis control [6–8], telemanipulation of robots [9], etc. Since different gestures are the result of the cooperative movement of several muscles, a machine-learning algorithm can be trained to learn and predict the gestures based on those muscle co-activation patterns.

Regarding our concept, three processings steps are required, namely: *preprocessing, feature extraction* and *classification*. The first two phases are similar to the ones seen in the previous section except that the encoding of only one hand gesture is required rather encoding a combination of gestures. The generated vector corresponds to a code composed of 8 binary values. The conversion process of each EMG signal into a binary code has been explained in the previous section. Finally, the classification consists of assigning a class label to features in the vector. In other word, the linear/non-linear classifiers assign to the vector the gestures that most probably belong to. Based on the literature, different classifiers have been proposed [10, 2], including Euclidean Distance, Logic Regression, k-Nearest Neighbors (kNN), Hidden Markov Model, Artificial Neural Network (ANN), Support Vector Machine (SVM), Linear discriminant Analysis (LDA) (see Fig. 7).

Within our context, one can consider two protocols for gesture recognition. The first protocol (one gesture against one subject) consists of detecting the gesture of one subject for the training and testing data sets of the *same* subject. Whereas, the second one (one gesture against M subjects) identifies the gesture of one subject for the training and testing data sets of *several* subjects.

Subjects achieve multiple gestures such as: fingers spread, fist, double tap, hand in, hand out, hand down and three. Each gesture is repeated N times by the same subject.

4.1 *Gesture Recognition for One Person in One Person Gestures Dataset (One Gesture Against One Person)*

The protocol is defined such that only one subject performs different sets of gestures. Each gesture is repeated several times.

Each single gesture registration generates a [L × N] matrix, where L is the registration length and N is the number of channels. This latter is transformed into binary codes of size L × N as explained in the previous section. For M number of gestures repeated T times, a [T × M × L × N] dataset matrix is collected.

4.2 Gesture Recognition for One Person in Several Persons Gestures Dataset (One Gesture Against n Persons)

The protocol is designed such that different subjects S performed different gestures. Each subject performs each gesture with repetitions. The challenging task is to identify the gesture of the associated subject.

Each gesture of each subject is transformed into binary code of size L × N. Within our context, it is considered M gestures repeated T times. So, we obtain a [T × M × L × N] dataset matrix for each subject. Consequently, the full dataset matrix of all subjects is S × [T × M × L × N].

4.3 Classification Algorithm

The classification algorithm is a key element of the EMG gesture recognition. In fact, its reliability and recognition efficiency have a great effect on the overall performance of the whole system. Recently, several algorithms have been proposed to serve certain scenarios [11, 12] like Linear Discriminant Analysis (LDA) [13–15] and SVM [16, 17]. In this chapter, we describe these four popular classical classification approaches that can be used to learn binary codes generated from EMG signals.

4.3.1 Linear Discriminant Analysis

LDA classifier consists of computing linear discriminant functions in order to select the maximum one as the classification rule. It aims to find a linear combination of the features in order to combine them into a single value. Its performance relies on how the features are combined. In an ideal way, features should be combined in such a way that if the classes are projected onto the axis they are seen as clearly separated tight clusters.

LDA is one of the most robust classifiers and it is good in reducing the dimensionality of features without separating the classes [18]. It showed high classification performance in several research works [18, 19] using a 10-fold cross validation in classification. Features were divided into 10 parts and each part takes turns to test. At the same time, the remainder used for training. In recent research work, Too et al.

demonstrated that the frequency domain features achieve a highest accuracy in LDA comparing with time domain features [20].

4.3.2 Support Vector Machine

Support Vector Machine is a discriminative classifier that separate hyperplane. Classification decision boundaries are identified by maximising a margin distance of training data. The training examples closest to the optimal hyperplane are called support vectors. Recently, Yamaba et al. [17] used an SVM classifier to identify gestures by s-EMG signals.

5 Conclusion

The purpose of this chapter is to highlight some concepts which consists in using a multi-channel EMG device in biometrics and hand gesture recognition. When considering biometrics in verification mode, the signals recorded through Myo device represent the contraction of the muscles involved in each hand gesture, through eight EMG signals. Once encoded, binary password can be generated to access any resource. The multi-channel EMG acquisition overcome the limitations of a single channel.

Even if the performance of such system has not been evaluated in this chapter, it is obvious that this aspects depend on the size of the database used for the training, including the number users, but also on the way the Myo armband is worn (i.e. condition of acquisition). In other words, in order to ensure an effective accuracy, it is important to make sure that the sensors are worn at the same position (landmarks may be required) for each acquisition session. Otherwise, the captured signals will wrongly be matched.

Another point that should be highlighted is the importance of using Machine-learning in verification mode, but also for gesture recognition. It is obvious that if the training is properly achieved, the performances increase.

As we will see in the next chapter, the idea of using a Multi-channel EMG biometics can be extended by using a High-Density sEMG acquisition. It may be another promising concept.

References

1. http://blog.thalmic.com/, 3 Feb 2018
2. Chowdhury, R.H., Reaz, M.B., Ali, M.A., Bakar, A.A., Chellappan, K., Chang, T.G.: Surface electromyography signal processing and classification techniques. Sensors (Basel) **13**, 12431–12466 (2013)

3. Zecca, M., Micera, S., Carrozza, M.C., Dario, P.: Control of multifunctional prosthetic hands by processing the electromyographic signal. Crit. Rev. Biomed. Eng. **30**, 459–485 (2002)
4. Zhang, X., Zhou, P.: High-density myoelectric pattern recognition toward improved stroke rehabilitation. IEEE Trans. Biomed. Eng. **59**, 1649–1657 (2012)
5. Liu, J., Zhou, P.: A novel myoelectric pattern recognition strategy for hand function restoration after incomplete cervical spinal cord injury. IEEE Trans. Neural Syst. Rehabil. Eng. 96–103 (2013)
6. Atzori, M., Müller, H.: Control capabilities of myoelectric robotic prostheses by hand amputees: a scientific research and market overview. Front. Syst. Neurosci. **9**, 162 (2015)
7. Ma, J., Thakor, N.V., Matsuno, F.: Hand and wrist movement control of myoelectric prosthesis based on synergy. IEEE Trans. Hum. Mach. Syst. **45**, 74–83 (2015)
8. Rasouli, M. Ghosh, R., Lee, W.W., Thakor, N.V., Kukreja, S.: Stable force-myographic control of a prosthetic hand using incremental learning. In: Proceedings of the 37th Annual International Conference IEEE Engineering in Medicine Biology And Society, IEEE (2015), pp. 4828–4831 (2015)
9. Jiang, H., Duerstock, B.S., Wachs, J.P.: A machine vision-based gestural interface for people with upper extremity physical impairments. IEEE Trans. Syst. Man Cybern. Syst. **44**, 630–641 (2014)
10. Riillo, F., Quitadamo, L.R., Cavrini, F., Gruppioni, E., Pinto, C.A., Pastò, N.C., Sbernini, L., Albero, L., Saggio, G.: Optimization of EMG-based hand gesture recognition: supervised versus unsupervised data preprocessing on healthy subjects and transradial amputees. Biomed. Signal Process. Control **14**, 117–125 (2014)
11. Ahsan, M.R., Ibrahimy, M.I., Khalifa, O.O.: EMG signal classification for human computer interaction: a review. Eur. J. Sci. Res. **33**(3), 480–501 (2009)
12. Kim, K.S., Choi, H.H., Moon, C.S., Mun, C.W.: Comparison of *k*-nearest neighbor, quadratic discriminant and linear discriminant analysis in classification of electromyogram signals based on the wrist-motion directions. Curr. Appl. Phys. **11**(3), 740–745 (2011)
13. Zhang, H., Zhao, Y., Yao, F., Xu, L., Shang, P., Li, G.: An adaptation strategy of using LDA classifier for EMG pattern recognition. In: Proceedings of the Annual International Conference IEEE Engineering in Medicine Biology and Society, IEEE, pp. 4267–4270 (2013)
14. Amsuss, S., Goebel, P.M., Jiang, N., Graimann, B., Paredes, L., Farina, D.: Self-correcting pattern recognition system of surface EMG signals for upper limb prosthesis control. IEEE Trans. Biomed. Eng. **61**, 1167–1176 (2014)
15. Jiang, X., Merhi, L.-K., Xiao, Z.G., Menon, C.: Exploration of force MyoGraphy and surface electromyography in hand gesture classification. Med. Eng. Phys. **41**, 63–73 (2017)
16. Tavakoli, M., Benussi, C., Lopes, P., Osorio, A.: Robust hand gesture recognition with a double channel surface EMG wearable armband and SVM classifier. Biomed. Signal Process. Control **46**, 121–130 (2018)
17. Yamaba, H., Kurogi, T., Aburada, A., Kubota, S., Katayama, T., Park, M.: Naonobu Okazaki: on applying support vector machines to a user authentication method using surface electromyogram signals. Artif. Life Robot. **23**(1), 87–93 (2018)
18. Zhang, D., Zhao, X., Han, J., Zhao, Y.: A comparative study on PCA and LDA based EMG pattern recognition for anthropomorphic robotic hand. In: IEEE International Conference on Robotics and Automation (ICRA), pp. 4850–4855 (2014)
19. Adewuyi, A.A., Hargrove, L.J., Kuiken, T.A.: Evaluating EMG feature and classifier selection for application to partial-hand prosthesis control. Front. Neurorobotics **10** (2016)
20. Too, J., Abdullah, A.R., Zawawi, T.N., Saad, N.M., Musa, H.: Classification of EMG signal based on time domain and frequency domain features. Int. J. Hum. Technol. Interact. **1**, 25–29 (2017)

Towards High Density sEMG (HD-sEMG) Acquisition Approach for Biometrics Applications

Mariam Al Harrach, Sofiane Boudaoud and Amine Nait-ali

Abstract This is the third chapter of this book dedicated to EMG biometrics modality. The purpose is to highlight a Multi-Channel technique based on a High Density sEMG (HD-sEMG) acquisition. In fact, HD-sEMG recording systems can be used to overcome the limitation of classical bipolar and monopolar sEMG recording systems. Consequently, in the considered concept, HD-sEMG system generates 64 EMG signals from which an EMG image is constructed and processed. Thereupon, it will be explained how one can deploy this technique in a biometric scheme.

1 Introduction

From the previous chapters, we addressed the concept of using surface EMG signals as an emerging biometric modality to verify person's identity. Performances of such biometric systems can obviously be affected by acquisition conditions. In fact, a single channel can capture the activity of a muscle at a specific location of the muscle. However, a multi-channel device allows a global representation of the muscle activity. For instance, Myo device (seen in the previous chapter) includes eight sensors. This type of acquisition is effective, but can probably be improved by increasing the spatial resolution by increasing the number of sensors around a specific surface. This concept is inspired from an existing solution employed basically in the medical field, called High Density sEMG technique.

Consequently, we will be discussing in this chapter, the idea of using High Density sEMG (HD-sEMG) technique to improve EMG based biometrics approaches. In fact, HD-sEMG recording systems have demonstrated their ability to overcome the limitation of classical bipolar and monopolar sEMG recording techniques. There-

M. Al Harrach
Polytech Angers, LARIS, 49100 Angers, France

S. Boudaoud
Laboratoire BMBI Compiègne, Université de Technologie de Compiègne, Compiègne, France

A. Nait-ali (✉)
Université Paris-Est, LISSI, UPEC, 94400 Vitry sur Seine, France
e-mail: naitali@u-pec.fr

A. Nait-ali (ed.), *Hidden Biometrics*, Series in BioEngineering,
https://doi.org/10.1007/978-981-13-0956-4_6

upon, we will try to explain how using this technique can be deployed in a generic biometric scheme. The main addition to the sEMG method will be the data fusion step wherein we will combine the information from 64 sEMG electrodes in order to extract the parameters that best describe each individual. This chapter is organized as fellows: in the following section, the limitation of classical EMG acquisition is presented. Afterwards, the HD-sEMG is described, then considered within a protocol of biometric application. Some preliminary results are then presented at the end of this chapter.

2 Limitation of Classical Surface EMG Recording Techniques

There are several elements that affect the sEMG signal. These elements not only disrupt the amplitude but also the spectral content of the sEMG [1]. A few of these factors are [1]: the electrodes configuration (size, alignme, etc.), the electrode position on the surface of the muscle, spatial filtering effect of the electrode, detection volume and muscle anatomy (motor unit number, motor unit type distribution, conduction velocity, etc.). All these factors can alter the sEMG signal properties. However, in this claim we will be focusing on the extrinsic causative aspects that are related to the electrode structures and their position on the skin.

The conventional sEMG recording techniques suffer from many limitations. One of the main limitations of the classic sEMG techniques is the deficient spatial resolution [2]. This can be explained by the difficulty of extracting the sources of activity from the sEMG signal. Therefore, it gives information about the activity of a group of MUs but not a single MU. Notting that obtaining information about the activity of a single MU is substantially beneficial in many domains such as therapy and neuromuscular disorder diagnosis and evaluation [2]. In addition, the sEMG signal collected from a certain location with respect to muscle using a monopolar or a bipolar configuration is not necessarily representative of the activity of the whole muscle [3]. Other limitation and a major problem of the conventional sEMG recording would be the electrode placement problem [3]. This parameter has direct influence on the precision of the latter applications like force estimation [4] for example. Thus, the placement of the electrodes with respect to the active MUs should be as constant as possible between subjects and recordings [3]. The recommendations for sensor placement for sEMG recordings are reported by Hermens et al. [5]. Finally, we have the problem of destructive superposition of the APs from different MUs due to their biphasic and triphasic nature which is known as auto-cancellation phenomenon [6]. All the aforementioned limitations could be overcome by using the HD-sEMG technique.

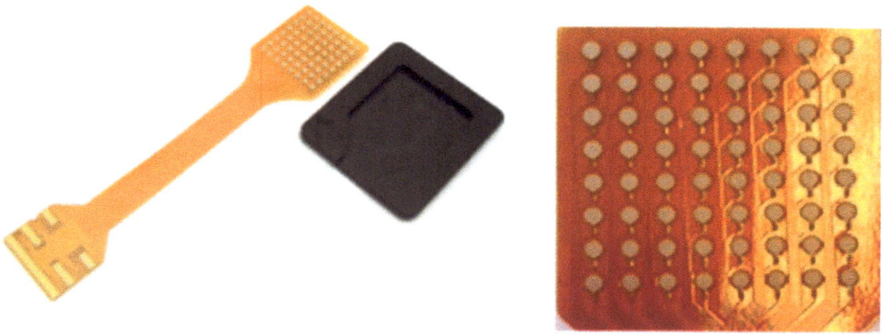

Fig. 1 An example of a HD-sEMG grid (TMSi property)

3 What Is HD-sEMG?

This novel sEMG recording technique is based on a multi-channel electrode recording system. It is composed of a two dimensional array of electrodes called a HD-sEMG grid (see Fig. 1) that is characterized by being a spatiotemporal variant of the usual single channel sEMG techniques [7]. The HD-sEMG technique overtops the classical singular and bipolar recordings by taking into account aspects of spatial distribution of electrical potentials [8] and by overcoming their basic limitations such as the electrode placement problem, the representativity of the recorded signal, the auto-cancelation phenomenon and the spatial resolution. It was also demonstrated through some researchers that it can improve force estimation [6, 8] and thereby prosthesis control [9]. The HD-sEMG recording technique has become widespread in the last few years, which is due to multiple reasons. Therefore, the HD-sEMG found its way into many applications due to the fact that it provides rich information about the muscle physiology and anatomy such as the MUs number [10] and properties [10] and the neural command by HD-sEMG decomposition [11]. In addition, the HD-sEMG found its way into clinical applications such as fatigue studies, neurogenic changes, myopathies and positive muscle phenomena [12]. Thus, becoming a valuable clinical diagnostic tool. Not to mention, the improvement that this novel technique has provided in force estimation [6].

4 HD-sEMG for Biometrics Schemes: Feasibility

In this section, we will describe a feasibility study using the HD-sEMG as a biometric modality in a generic identification scheme. Thence, we will basically present the experimental context of the data acquisition as described in the experimental protocol [13]. Afterwards, we will present the data fusion method based on image segmentation based on the algorithm used presented in [14, 15]. Finally, extracted features will be injected in an Artificial Neural Network (ANN) in order to achieve a classification for subject identification.

4.1 Experimental Protocol and Data Acquisition

In this section, we will report the details of the experimental protocol used in order to collect the HD-sEMG signals used for the study.

For this preliminary experiment, a small database has been created wherein 7 healthy male subjects without any musculoskeletal pathology (mean ± std, age: 24.6 ± 1.1 years; stature: 175.7 ± 3.1 cm; body mass: 70.9 ± 5.5 kg) participated in the study respecting the declaration of Helsinki. All participants provided their written consent before participating in the study and did not relate any history of muscular or neurophysiological problem at the upper limbs.

The sEMG signals of the two dimensional adhesive array consisting of 64 electrodes of circular shape with an inter-electrode distance of 4 mm and an electrode diameter of 1.5 mm were recorded from the Biceps Brachii (BB), during isometric voluntary contractions. The reference electrode and the ground were placed on the elbow bone and around the wrist respectively [13].

The monopolar sEMG signals were acquired using TMSI REFA 136 acquisition system (Twente Medical Systems International, the Netherlands) with the dedicated software (Polybench). The global force signal was measured simultaneously with the sEMG signals using a strain gauge (Celtron STC SS S-beam, Stainless Steel, IP67), and displayed in real-time on an oscilloscope as visual feedback to the subjects. This protocol allowed us to obtain, for each subject, 64 signals of 5 s.

Subjects were seated upright in a chair with the forearm in a horizontal position, and the elbow angle fixed at 90°. After skin preparation (shaving, abrading and cleaning) the HD-sEMG grid of 64 electrodes was placed either proximally (~1.5 cm) from the middle of the BB muscle according to the SENIAM recommendations [13]. Isometric elbow flexion force was recorded synchronously with the HD-sEMG signals. The protocol started by measuring the MVC. This was done in three trials, each trial lasted 3 s with a 2 min rest period in between in order to avoid fatigue accumulation. The MVC was chosen as reference for sub-level contractions. Three trials of isometric contractions were recorded at 20, 30, 40, 50, 60 and 70% MVC by providing a target to the subject with 1.5 min rest in between. The target signal was displayed in real time on the oscilloscope along with the subject's instantaneous contraction level as a feedback. These force plateaus lasted 5 s wherein the subject was trained to keep the force as constant as possible by watching the feedback signal on the oscilloscope.

4.2 Data Fusion Technique

The HD-sEMG technique can be considered as an imaging tool [16] in complement to sonography and Magnetic Resonance Imaging (MRI) for assessing muscle systems. Therefore, we used an automatic and personalized channel selection method based on image segmentation that was previously used and proven effective [14]. This method is obtained using a previously validated HD-sEMG image segmentation method, the Watershed (WS) algorithm [15]. This approach is a non-parametric unsupervised segmentation method that selects an optimal threshold based on discriminant analysis to maximize the separability of the classes in gray levels [17].

As mentioned before, using traditional bipolar electrodes cannot permit a loyal illustration of the entire muscle activity, not to mention the positioning problem. The use of a two-dimensional electrode network can improve the efficiency of later applications [6]. In fact, this bi-dimensional recording technique allows us to obtain a map of the electrical activity present on the skin surface. This map can be considered as a picture, wherein the pixels represent a certain parameter values computed over the signal of the corresponding electrode, such as an amplitude descriptor, computed for a certain time span. However, the whole picture is not always meaningful; the area of interest may change with the contraction level produced or by the subject generating this force. Accordingly, it is important to develop an effective technique to draw out the region of interest from these maps and subsequently fusion the information contained within. Such a technique is called image segmentation. Many segmentation algorithms exist in the literature. A recent work by Vieira et al. [15] presented an automatic segmentation of sEMG signals in the objective of a better estimation of the neuromuscular activity. They used a well-known segmentation algorithm, the WS, and added some steps in order to enhance the segmentation results [15]. Afterwards, they compared this approach with other segmentation techniques (h-dome and k-means segmentation algorithms) and found that the WS technique is more robust and accurate compared to the others [15].

The principle of this WS transformation is based on the idea that any grayscale image can be considered as a topographic. And if this surface was flooded from its minima without the merging of waters coming from different sources, this image could therefore be divided into two different sets: the catchment basins and the WS lines [15]. The algorithm consists on identifying the location of ridges (WSs) in the grayscale image and labeling the catchment basins with different numbers. This algorithm is presented below.

The Watershed Algorithm

(a) A set of markers, pixels where the flooding shall start, are chosen. Each is given a different label.
(b) The neighboring pixels of each marked area are inserted into a priority queue with a priority level corresponding to the gradient magnitude of the pixel.

(c) The pixel with the lowest priority level is extracted from the priority queue. If the neighbors of the extracted pixel that have already been labeled all have the same label, then the pixel is labeled with their label. All non-marked neighbors that are not yet in the priority queue are put into the priority queue.
(d) Redo step (c) until the priority queue is empty. The non-labeled pixels are the WS lines.

4.3 Extraction of Parameters

The thresholding was done on monopolar Root Mean Square (RMS) value maps for the studied contraction levels (20–70% MVC). We chose the RMS value instead of the Averaged Rectified Value (ARV) based on its robustness to spatial aliasing.

RMS map generation
The HD-sEMG signals present a grid of small electrodes where each electrode can be considered as a pixel of the EMG map or image. The RMS map was computed for the EMG signals corresponding to a constant isometric force of 5 s using Eq. (1) with an epoch of 1 s (2000 samples).

$$RMS(i,j) = \frac{1}{N}\sum_{n=1}^{N}\sqrt{\left(\frac{1}{K}\sum_{k=1+(n-1)K}^{nK} EMG^2(i,j,k)\right)} \tag{1}$$

where i, j are the row and column numbers respectively, K is the epoch length and N is the number of sample in each epoch. In this study K and N were equal to 2000 and 5 respectively. An example of the computed RMS image in grayscale is presented in Fig. 2.

RMS map segmentation
In our case pixels with high RMS values would be perceived as elevated surfaces. Therefore, applying the WS algorithm directly to the RMS map leads to WS line comprising of pixels representing high neuromuscular activity (high gray intensity) which is not desired (see Fig. 3). Instead, WS algorithm will be applied on the gradient of the RMS map (G) which provides the degree of change in grayscale image [15]. The gradient map is computed as follow:

$$G_x(mn) = F^{-1}\left(S^T \sum_{m=1}^{M}\sum_{n=1}^{N} RMS(m,n)e^{-j2\Pi\left(\frac{k_x m}{M}+\frac{k_y m}{N}\right)}\right)$$

$$G_y(mn) = F^{-1}\left(S \sum_{m=1}^{M}\sum_{n=1}^{N} RMS(m,n)e^{-j2\Pi\left(\frac{k_x m}{M}+\frac{k_y m}{N}\right)}\right)$$

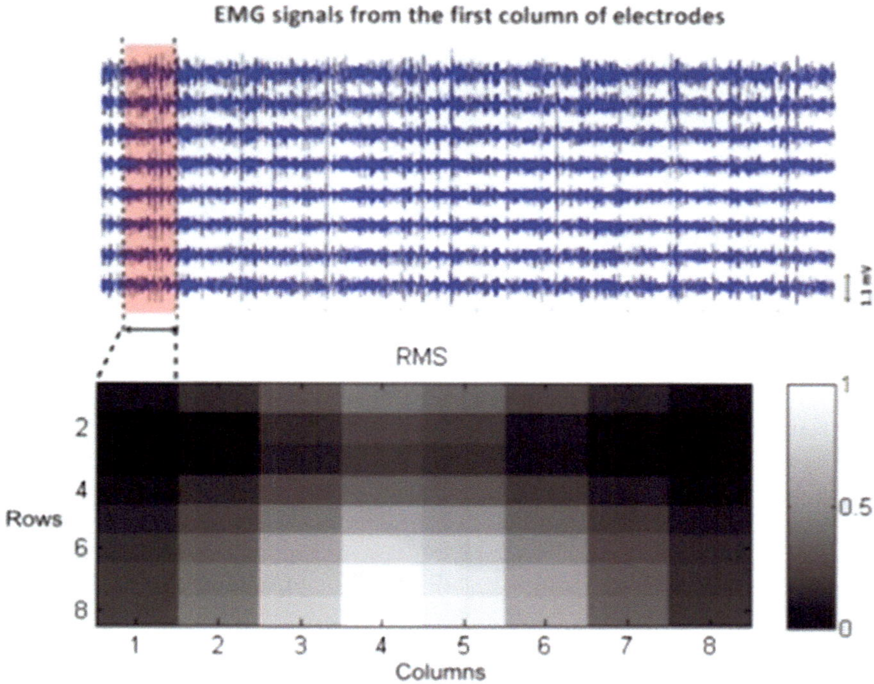

Fig. 2 The grayscale image (bottom panel) created with the RMS values computed for an experimental EMG recorded from the BB during isometric flexion at 30% MVC and the corresponding EMG signals for the first column (upper panel)

$$G = \sqrt{G_x^2 + G_y^2}$$

where the inverse Fourier is transform operator, and are the spatial frequencies, M and N the number of rows and columns in the HD-sEMG grid and S the bi-dimensional Fourier transform of the zero-padded Sobel operator presented below:

$$S = \begin{matrix} +1 & +2 & +1 \\ 0 & 0 & 0 \\ -1 & -2 & -1 \end{matrix}$$

Since the number of clusters obtained by the WS segmentation depends on the number of regional minima in the gradient (G), the problem of over segmentation could be confronted by smoothing sharp transitions in G using image opening and image closing operations [15]. These operations are operators from mathematical morphology derived from the fundamental operations of erosion and dilation [18]. Opening and closing G using the structural element v are defined by [18]:

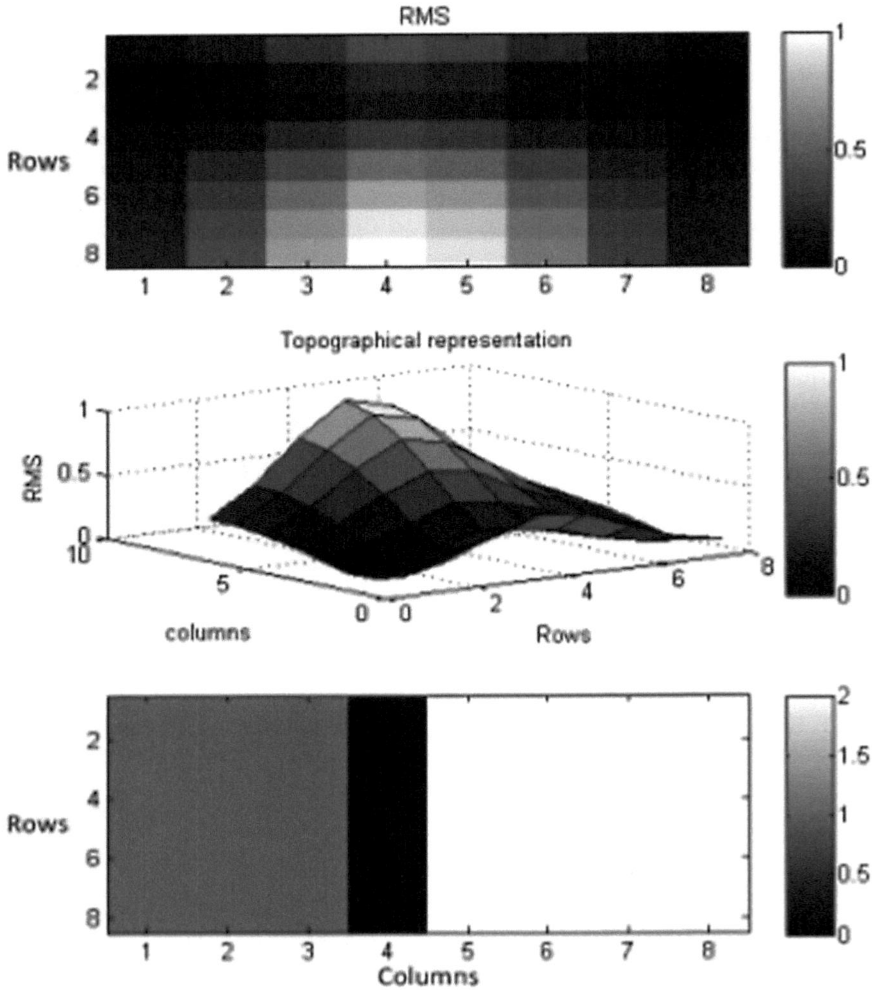

Fig. 3 Schematic of the WS segmentation (bottom) applied to the grayscale image (Top) obtained from the RMS map of simulated EMG signals with its topographical representation (middle)

$$G \circ v = (G \ominus v) \oplus v$$

$$G \cdot v = (G \oplus v) \ominus v$$

where the $\bullet$ and $\circ$ indicate the opening and closing respectively, and are the Minkowski operators for addition and difference defined by [18]:

$$(G \oplus v)(p) = max_{z \subset D_v}\{G(p+z)\}$$

$$(G\ominus v)(p) = min_{z\subset D_v}\{G(p+z)\}$$

where is the Domain of the structuring element v, which was chosen as a square grid (3 × 3) of ones (i.e. z • [1, 0, 1] × [1, 0, 1]).

Figure 4 illustrates the clustering result of the WS algorithm applied to the gradient and the smoothed gradient of the same RMS map presented in Fig. 3.

Also, we have to mention that we used the 4-type connectivity algorithm. This means that each pixel has four connected neighbors (top, bottom, right and left). The final step of this method is the thresholding which consists on removing further pixels or channels by applying a threshold equal to some percentage of the maximum RMS value in the RMS map (70% in our case).

After segmentation of the RMS maps we obtain a number of channels specific to each subject. These channels are the channels that are most representative of the muscle activation. Thus, we will compute the different parameters for these channels and then we will average the values to obtain one value per parameters for each subject at each contraction level and for 5 windows (window of 1s).

Thus, we started by estimation of the DSP using the periodogram method and then we calculated the spectral parameters: Power, median frequency, mean frequency, peak frequency, deciles and HOS.

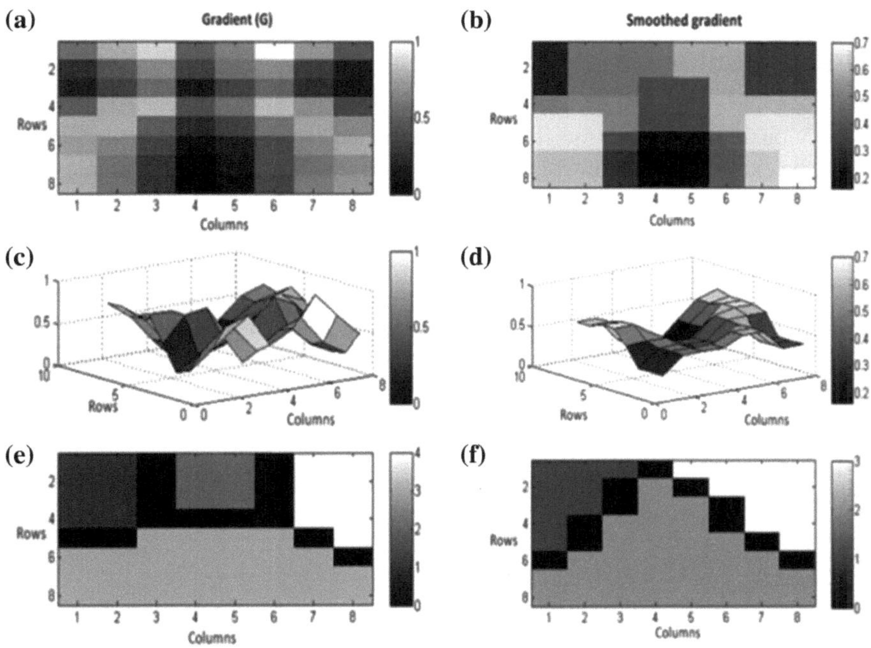

Fig. 4 Schematic of the WS segmentation results (**e**) and (**f**) applied to: **a** the gradient (G) of the image and **b** the smoothed gradient obtained by opening and closing techniques respectively. With their corresponding topographical representation (**c**) and (**d**)

High Order Statistics (HOS)

The HOS parameters, the skewness (asymmetry) and the kurtosis (flatness), track indirectly the PDF shape modifications of the sEMG [19]. We recall briefly the definitions of both normalized parameters in the following equations for the sEMG amplitude:

$$skew(sEMG) = \frac{E\left[(sEMG - \mu)^3\right]}{\left(E\left[(sEMG - \mu)^2\right]\right)^{\frac{3}{2}}}$$

$$kur(sEMG) = \frac{E\left[(sEMG - \mu)^4\right]}{\left(E\left[(sEMG - \mu)^2\right]\right)^2} - 3$$

where E(.) is defined as the expectation operator, and μ is the mean value of the sEMG signal amplitude. It is important to note, that these high moments are invariant to the mean value and variance variability. These parameters were computed for 5 s of constant force plateaus by averaging the values obtained over ten epochs of 0.5 s (sampling frequency, fe, fixed at 2048 Hz) in order to minimize the bias of the used HOS estimators that employ empirical formula.

5 Classification Using Artificial Neural Network (ANN) and Results

As previously mentioned, the signals were obtained from 7 healthy male subjects. Thus, the experimental protocol allowed us to have a database of 210 samples of 17 parameters.

For the classification, we used a Radial Basis Function (RBF) neural network with 17 inputs corresponding to the computed parameters and 7 outputs corresponding to the 7 individuals. As for the hidden layer, we tried different number of neurons. The preliminary results of the identification are presented in Table 1.

By observing Table 1, we can see that we have the optimal result for 150 hidden neurons.

Table 1 The performance of the classification

Number of hidden neurons	MSE
10	0.122
50	0.089
100	0.055
150	0.020
200	1.386

6 Conclusion

This preliminary study of HD-sEMG based biometrics allowed us to confirm our initial hypothesis that this technique can be beneficial in the sEMG based identification. In fact, HD-sEMG adds spatial information to the temporal propagation captured by classical sEMG methods. Thus, it adds more definition and subject specificity to the extracted parameters especially the Probability Density Function (PDF) shape descriptors, the HOS. Along these lines, it has been shown in a recent study [19] that by studying the HOS profile variation following the contraction level increase; we can obtain information about the MU type distribution in the muscle. In addition, it has also been shown that by using these variations of HOS, the kurtosis and the skewness, we can achieve a classification into group depending on the percentage of MU types in their muscle. We have to note here that this parameter, the MU type distribution, cannot be measured using noninvasive procedures. Therefore, we think that the combination of HD-sEMG of 64 signals with the data fusion step allows us to add uniqueness to each parameter extracted from an individual which permits to improve the identification process.

References

1. De Luca, C.J.: The use of surface electromyography in biomechanics. J. Appl. Biomech **13**, 135–163 (1997)
2. Merletti, R., Parker, P.A.: Electromyography: physiology, engineering, and non-invasive applications. IEEE Press, Piscataway; Wiley-Interscience, Hoboken (2004)
3. Staudenmann, D., Roeleveld, K., Stegeman, D.F., van Dieen, J.H.: Methodological aspects of SEMG recordings for force estimation—a tutorial and review. J. Electromyogr. Kinesiol. **20**(3), 375–387 (2010)
4. Disselhorst-Klug, C., Schmitz-Rode, T., Rau, G.: Surface electromyography and muscle force: limits in sEMG–force relationship and new approaches for applications. Clin. Biomech. **24**(3), 225–235 (2009)
5. Hermens, H.J., Freriks, B., Disselhorst-Klug, C., Rau, G.: Development of recommendations for SEMG sensors and sensor placement procedures. J. Electromyogr. Kinesiol. **10**(5), 361–374 (2000)
6. Staudenmann, D., Kingma, I., Daffertshofer, A., Stegeman, D.F., van Dieen, J.H.: Improving EMG-based muscle force estimation by using a high-density EMG grid and principal component analysis. IEEE Trans. Biomed. Eng. **53**(4), 712–719 (2006)
7. Merletti, R., Afsharipour, B., Piervirgili, G.: High density surface EMG technology. In: Converging Clinical and Engineering Research on Neurorehabilitation, no. 1 in Biosystems & Biorobotics, pp. 1205–1209. Springer Heideblerg, Berlin
8. Rojas-Martinez, M., Mananas, M.A., Alonso, J.F.: High-density surface EMG maps from upper-arm and forearm muscles. J. NeuroEng. Rehab. **9** (2012)
9. Daley, H., Englehart, K., Hargrove, L., Kuruganti, U.: High density electromyography data of normally limbed and transradial amputee subjects for multifunction prosthetic control. J. Electromyogr. Kinesiol. **22**(3), 478–484 (2012)
10. van Dijk, J.P., Blok, J.H., Lapatki, B.G., van Schaik, I.N., Zwarts, M.J., Stegeman, D.F.: Motor unit number estimation using high-density surface electromyography. Clin. Neurophysiol. **119**(1), 33–42 (2008)

11. Kleine, B.U., van Dijk, J.P., Lapatki, B.G., Zwarts, M.J., Stegeman, D.F.: Using twodimensional spatial information in decomposition of surface EMG signals. J. Electromyogr. Kinesiol. **17**(5), 535–548 (2007)
12. Drost, G., Stegeman, D.F., van Engelen, B.G., Zwarts, M.J.: Clinical applications of highdensity surface emg: a systematic review. J. Electromyogr. Kinesiol. **16**(6), 586–602 (2006)
13. Al Harrach, M., Boudaoud, S., Hassan, M., Ayachi, F.S., Gamet, D., Grosset, J.F., Marin, : Denoising of HD-sEMG signals using canonical correlation analysis. Med. Biol. Eng. Comput. 1–14 (2016)
14. Al Harrach, M., Carriou, V., Boudaoud, S., Laforet, J., Marin F.: Analysis of the sEMG/force relationship using HD-sEMG technique and data fusion: a simulation study. Comput. Biol. Med. 34–47 (2017)
15. Vieira, T.M., Merletti, R., Mesin, L.: Automatic segmentation of surface EMG images: Improving the estimation of neuromuscular activity. J. Biomech. **43**(11), 2149–2158 (2010)
16. Merletti, R., Holobar, A., Farina, D.: Analysis of motor units with high-density surface electromyography. J. Electromyogr. Kinesiol. **18**(6), 879–890 (2008)
17. Sahoo, P.K., Soltani, S., Wong, A.K.C.: A survey of thresholding techniques. Comput. Vision Graph. Image Process. **41**, 233–260 (1988)
18. Heijmans, H.J.A.M.: Mathematical morphology: a modern approach in image processing based on algebra and geometry. SIAM Rev. **37**(1), 1–36 (1995)
19. Al Harrach, M., Boudaoud, S., Carriou, V., Laforet, J., Marin, F.: Investigation of the HD-sEMG probability density function shapes with varying muscle force using data fusion and shape descriptors. Comput. Biol. Med. 44–58 (2017)

Age Estimation Using Sound Stimulation as a Hidden Biometrics Approach

Muhammad Ilyas, Alice Othmani and Amine Nait-ali

Abstract In this chapter, it will be introduced a new hidden biometrics approach of age estimation requiring the stimulation of the auditory system by an acoustical modulated sine wave signal. After a quick review on different common approaches used in the field of age estimation, and after presenting some generalities on the auditory system, age estimation and age classification protocols will be considered. This chapter describes also the concept of a simple identification/verification, as an application.

1 Introduction

In the present era, the technological advancement covered almost every facet of life. Biometrics is providing security to such environment to keep our privacy safe. Biometrics is a set of unique human physical and behavioral features that are used for identification and verification.

Human age is an important factor among biometric features as an effective signature. It is the center of attention for the scientific communities such as pattern recognition, computer vision, and biometrics. The need for security and smart system drove the interest of researchers for human age estimation in forensic science analysis and human-computer interaction. Hence, the study of human age for forensic and security tools is used by considering the physical and physiological appearances or behavioral features of a human.

Most of the studies are focused on age estimation, mainly, either from the face or voice analysis [1]. For efficient human age estimation physical, physiological and behavioral biometric features have been used.

In this chapter, we introduce a new hidden biometrics approach of age estimation requiring the stimulation of the auditory system by an acoustical modulated sine wave signal. After a quick review on different common approaches used in the field of age estimation, and after presenting some generalities on the auditory system, Sect. 4

M. Ilyas · A. Othmani · A. Nait-ali (✉)
Université Paris-Est, LISSI, UPEC, 94400 Vitry sur Seine, France
e-mail: naitali@u-pec.fr

A. Nait-ali (ed.), *Hidden Biometrics*, Series in BioEngineering,
https://doi.org/10.1007/978-981-13-0956-4_7

will describe age estimation and age classification protocols. Finally, the concept of a simple identification/verification application is highlighted.

2 Review

In this section, different techniques and approaches of biometrics age estimation will be briefly evoked. In particular three main categories will be considered, namely: Physical Biometrics, Physiological Biometrics and Behavioral Biometrics.

Physical Biometrics for Age Estimation: Face is a physical biometric feature which is widely used for age estimation. Indeed, face conveying a significant amount of information such as the identity, ethnic group, gender, emotional state, and age. Despite the fact that the age of a person is an important attribute that determines the way we interact with a person, there has been less work focused on developing systems for automatic age estimation than for human identification from face photos [2]. From the technical point of view, different classifiers based on quadratic functions, shortest distance and neural network are performed [3] while for age classification, a dropout-SVM that avoids over-fitting is used [2]. While the error of estimation from face image was around five years [1, 2], it appears that human still achieves better results than machines on age estimation with an age estimation errors of 3.64 years [2].

Physiological Biometrics for Age Estimation: In forensics, DNA or dental traits is used to estimate the human age. The forensic individual age estimation with DNA have surged in the last five years. For forensic age estimation, morphological inspection or radiography was used before and later molecular approaches have been adopted. Blood samples are used for age estimation, but recently scientist is also using tissue samples for age estimation. The predictive accuracy has been increased the potential of the system and its accuracy with an error value of ± 4 years [4]. For children, the age is estimated through dental maturation while for adults, morphological and radiological techniques are used [5].

Behavioral Biometrics for Age Estimation: Voice is behavioral biometrics, used for identification and authentication. Basically, voice recognition systems analyze and extract features from the subjects speech signal. Voiceprints are used in many forensic applications to identify automatically the gender, age, and language of a subject. The automatic age is classified and estimated into a specific age group using parallel phone recognizer, dynamic Bayesian networks, linear prediction analysis, HMM-based classifiers [6] gaussian mixture models with MFCCs [7], Gaussian Mixture Model (GMM) supervectors and Support Vector Regression [8]. Automatic age estimation of a speaker is also proved possible with an absolute mean error which varies from 4.7 to 10 years [1] and classification accuracy around 50% [6].

Gait signatures are also considered as one of the best behavioral biometric features for human age estimation and classification. Using gait features, it is possible to recognize gender (males, females), even in the same age group. For gait-based age estimation, multi-label-guided (MLG) subspace is presented to correlate and characterize the gender and age of a person [9] and a baseline algorithm using Gaussian process regression with silhouette-based gait features is used [10]. The mean absolute error is from 6 to 8.2 years for the proposed approaches [9, 10].

Keystroke dynamics is also a perspective of age detection. With the help of keystroke dynamics, the exact age of a subject cannot be estimated, but it can be utilized to differentiate between kids and adults. A multi-layer perceptron shows that the age can be detected with a probability that it is far from the uniform distribution with a success rate of 72% for only two classes and 64% for four classes [11]. In previous work, a new approach to distinguish between child and adult is proposed, while the accuracy rates above 90% were achieved with a support vector machine or linear discriminant analysis classifier [12].

3 Human Auditory System: Generalities

The human auditory system is the sensory mechanism for the sense of hearing. It is having two sensory organs to hear the sound known as ears, which are connected to both the sides of the head and connected to brain stem through a brisk nerve fiber. The salient linking point of the brain is called brain stem which provides a bridge to the human nervous system to the other parts of the brain. The auditory information is processed through a temporal lobe called auditory cortex [1].

The human ear is extremely an intricate organ. This process is making furthermore difficulty when the information from the two ears is collectively transferred through a mystifying neural network called the brain. Here just a short story about the ear is presented, many elusive possessions and poorly understanding marvels are still found associated with human hearing.

Figure 1 explains the main processes and their structures that cover a normal human ear. The outer ear which is the combination of two portions, the observable flap of skin and cartilage attributed to the head side, and the canal of an ear, a tube

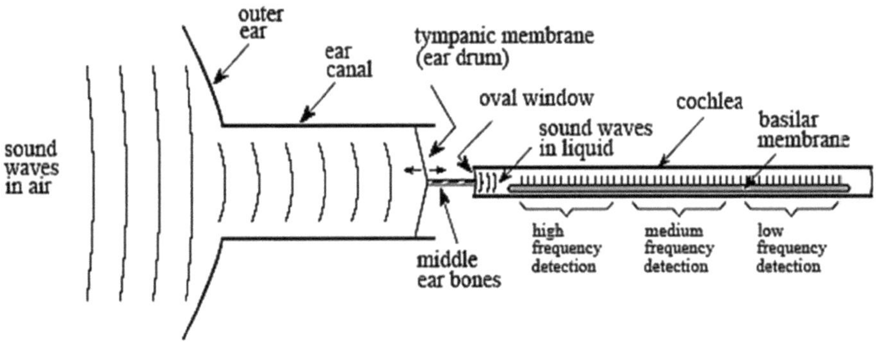

Fig. 1 Functional diagram of human ear [13]

nearby 0.5 cm in diameter ranging around 3 cm into the head region. These structures forward the surrounding effects to the middle and inner ear organs located securely inside the bone of the skull. Stretched transversely found at the ear canal end is a tinny sheet of tissue named eardrum (tympanic membrane).

It causes vibration when the Sound waves stick to it. The middle ear is a combination of small bones that forward the vibration of the eardrum to the cochlea (inner ear) to convert into neural impulses. The cochlea is a tube-like structure filled from liquid about 2 mm in width and 3 cm in length. Though revealed straight in Fig. 1, cochlea of the ear is curly in shape which looks like a small snail shell. In reality, the cochlea is derivative of the Greek word used for a snail.

At that time, when a wave of sound efforts to cross air state to liquid, so only a small portion of these waves are diffusing into surrounding whereas the remaining energy is reflected due to truncated mechanical impedance (low acoustic pressure and high particle velocity resulting from low density and high compressibility) of air as compared to liquid having high mechanical impedance. It is concluded that mechanical impedance difference mostly reflected the sounds waves in the air medium or fluid.

The network function of the middle ear increases a portion of the energy of the sound while entering the inner ear liquid medium. Mostly this impedance alteration is coming from the difference in zone amongst the eardrum (sound received from the air) and the oval window (see Fig. 1). The zone of the eardrum and oval window is roughly estimated as 60 mm^2 and 40 mm^2 respectively. So, the applied pressure is equal to force separated through a zone and this zone difference increases the pressure of sound wave approximately 15 times.

There is a basilar membrane inside the cochlea works like supporting building composed of usually 12,000 sensory cells which form the cochlear nerve. This membrane is present in a solid form adjacent to the oval window and turns elastic form towards the opposite end, permitting it to perform a role of the frequency spectrum analyzer. Because of exposure to the high-frequency signal, this membrane vibrated due to its rigid nature and exited the nerve cells to move near the oval window and vice versa. This association which makes fibers specific in the cochlear nerve as a result of to specific frequencies responses named place principle, which is well-preserved through the aural path mad about the brain.

Volley principle is another information encoding scheme in the hearing of human. And those electrical beats which are produced by nerve cells when transfer information is term as action potentials. Action potentials are generating in responses of every round of vibration when a nerve cell of the basilar membrane code auditory materials. E.g., a two hundred Hz of sound waves can be produced through a neuron generating an action potential of two hundred/sec at frequencies lower than 500 Hz. It is the determined fraction of brain neurons to produced action potentials. So, this problem is overwhelming through the human ear which permitting many nerve cells to perform this type of task. A given an example will clarify this statement. A three thousand Hz capacity might be signified by ten nerve cells consecutively fire 300/s. This calculation covers the assortment of volley principle up to four kHz, and above this figure, the place principle completely applies. Some Specific occurrences responses make a precise fiber in the cochlear nerve, which is known as place principle, preserved the auditory way to brain completely.

In Table 1, the interrelationship between perceived loudness and sound intensity is briefly explained. To indicate the sound intensity on a logarithmic scale is known as decibel SPL (Sound Power Level). 0 dB SPL has a sound wave power of 10–16 W/cm^2, about the lowest intensity of sound perceived by a human ear. A typical speech is about 60 dB SPL, while severe damage can occur to the ear at nearly 140 dB SPL.

The difference between the faintest and loudest sounds is that human can hear about 120 dB. A person can quickly detect the change in loudness of sound when the signal is modified by about 1 dB (a change of 12% in amplitude). While there are about 120 levels of loudness that a human can perceive from the low whisper to the thunder. The human ear has an amazing sensitivity while listening to the low-frequency sound, and the eardrum vibrates much lesser than the diameter of a molecule!

The range of human hearing is usually considered between 20 and 20,000 Hz, but the sensitivity to sound starts from 1000 to 4000 Hz. For example, a person can detect sound as lower as 0 dB SPL at 3000 Hz but needs 40 dB SPL at 100 Hz (an increased amplitude to 100). A person can judge that the two sounds are different if the frequencies differ by more then 0.3% at 3000 Hz and increases to 3% at 100 Hz. A good example is that adjacent keys of a piano differ by 6% in frequency.

Table 1 Sound intensity levels and intensities

W/cm^2	Decibel	Effects/examples
10^{-2}	140	Jet airplane at 30 m, severe pain, damage in seconds
10^{-3}	130	
10^{-4}	120	Loud rock concert, pneumatic chipper at 2 m, threshold of pain
10^{-5}	110	Damage from 30 min per day exposure
10^{-6}	100	Noisy factory, siren at 30 m; damage from 8 h per day exposure
10^{-7}	90	Inside a heavy truck; damage from prolonged exposure
10^{-8}	80	Loud radio, classroom lecture
10^{-9}	70	Noisy office, busy traffic
10^{-10}	60	Normal conversation
10^{-11}	50	Average office, soft music
10^{-12}	40	Average home
10^{-13}	30	Quiet home
10^{-14}	20	Whisper at 1 m distance
10^{-15}	10	Rustle of leaves
10^{-16}	0	Threshold of hearing at 1000 Hz
10^{-17}	10^{-1}	Weakest audible
10^{-18}	10^{-2}	

The first advantage of having two ears is the capability of identifying the direction of the sound. The human can detect the difference between two sources sound that is put as far as three degrees apart, about 10 m to the width of the person. The information about direction is received in two different ways. First, the frequencies above 1000 Hz are strongly tracked by the head such that the ear near to the sound source gets a stronger signal as compared to the ear on the other side of the head. The second sign to directionality is, that ear which is far from the sound source side hears the sound slightly later compare to the near ear, due to its more considerable distance from the sound source.

While human hearing can quickly determine the direction of a sound source and identification of the distance to the sound source is poor. The reason is that very few signs are available in a sound wave that can provide such information. Human hearing poorly perceives that low-frequency sounds are distant while the high-frequency sounds are nearby [13].

4 Age Estimation and Classification: A Global Approach

Human perception is the ability to interpret information that our senses receive from the environment. The auditory system is the sense that allows us to discern and locate sounds. The capability to receive and explicate these sounds that reaches the ears through sound waves transmitted through air or other means is called auditory perception. While at the age of 16 years, our highest audible frequency is nearly 180,00 Hz. At the age 30 years, it decreases to about 150,00 Hz. This means that as the age increases, the highest audible frequency decreases in parallel. Hence, there is a strong correlation between the age and the audible frequency [14–17]. Therefore, it leads to hearing loss that occurs over the time due to the damage of hair cells in our auditory system [18–21].

Considering the factor of hearing loss, age can be estimated using the auditory perception. Despite that in literature it is mentioned that human has the capability to hear a sound of 12 Hz under favorable conditions, and the commonly stated range of human hearing is from 20 to 20,000 Hz [14–16].

Human age estimation through auditory perception has been introduced for the first time [17]. To better study the correlation between age and auditory perception, two objectives have been presented:

1. Classification of the auditory perception responses into different age groups
2. Human age estimation based on the auditory perception responses.

4.1 Protocol

A dynamic frequency sound is generated to stimulate the human auditory system according to the following model:

$$x(t) = A_0 \sin(2\pi \, . \, \phi(t) \, . \, t)$$
$$\text{where } \phi(t) = \alpha \, . \, t + \phi_0$$

A_0 stands for sound amplitude, t stands for time, ϕ_0 is the initial frequency, and α stands for the increasing/decreasing value of frequency.

Four different protocols with the speaker (S) and headphone (H) are considered namely:

1. Bilateral Stimulation (S-BS),
2. Bilateral Stimulation (H-BS),
3. Right Stimulation (H-RS),
4. Left Stimulation (H-LS).

In the first protocol, the duration is set to $t_1 = 20$ s, $t_2 = 40$ s, and $t_3 = 60$ s, protocol allowed the system to define the appropriate duration for stimulation. In the remaining three protocols, headphones are used by considering Bilateral Stimulation (H-BS), Right Stimulation (H-RS), and Left Stimulation (H-LS).

Therefore, a subject conduct six experiments as shown in Fig. 2. For each protocol, the subject interacts with the system in real time. The sound is generated and the subject should react with a manual action to stop the system. The experiment is conducted conferring to two modes:

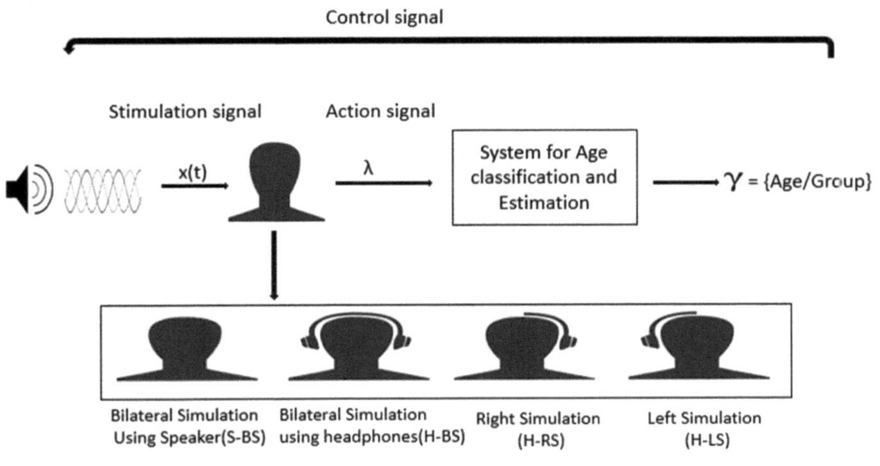

Fig. 2 Flow diagram of our proposed approach

- first mode: the sound is increased from the lower frequency (20 Hz) to a higher frequency (20,000 Hz). The subject stops the stimulation (e.g. keyboard action), once stop hearing,
- second mode: a second stimulation is generated automatically once the subject complete the first mode action. In this case, the generated sound frequency decreases from the higher frequency (20,000 Hz) to the lower frequency (20 Hz). The volunteer stops the stimulation once the sound is perceived.

4.2 Human Age Classification Using Auditory Perception

This section concerns the classification of the auditory perceived responses into different age groups. For classification supervised learning method is used based on Random Forest (RF). RF classifiers are widely used in the field of computer vision and pattern recognition applications because of there proven output.

RF was proposed for the first time by Leo Breiman in 2001 [22]. RF is a set of a large number of decision trees which are developed on bootstrap samples of the training data, selecting the variables subsets randomly. To determine a sample's class each tree of the forest votes and a majority vote for a class makes the final decision. For the age classification system, RF classifier is built with recommended values by Breiman. The number of decision trees is equal to 500 while the number of features is used to split the node in the decision tree growing process which is denoted by M_{try}. It is fixed to a value of 0.10, which is closer to the recommended value proposed by Breiman $M_{try} = \sqrt{p}$ where p is the feature vector size and its value is 3. A testified k-fold cross-validation technique is applied using RF classifiers with k = 10. Cross-validation is a strategy to quantify a predictive models by dividing the original data into a training set for model training, and a test set to evaluate [23–25]. The performance of the classifier can be enhanced by optimizing the parameters of the algorithm. Cross-validation performance is figured out and then the best performing parameters inside the loop could be selected. Then, classification with best parameters is applied to the original testing dataset [26, 27].

The classification performance of the auditory perception into three and five age groups is performed using RF classifiers is shown in Table 2. The phenomenon is repeated 100 times for ten-fold cross-validation. The best accuracy of 92% for three age groups is achieved for Bilateral Stimulation using speakers $t_1 = 20s$ dataset with RF classifier. For five age groups, the highest accuracy of classification of 86% also with the S-BS for $t_1 = 20s$ dataset and the RF classifier.

Table 2 Performances of the classification into three or five age groups using RF

Dataset	Random forest	
	Three age groups (%)	Five age groups (%)
S-BS(20 s)	92	86
S-BS(40 s)	91	84
S-BS(60 s)	86	80
H-BS	86	81
H-RS	86	80
H-LS	84	81

4.3 *Human Age Estimation Using Auditory Perception*

A regression model has been designed for human age estimation using Regression Forest (RF) as shown in Fig. 3.

Two types of techniques have been used: Regression Forests and Support Vector Regression. Regression Forests (RF) is a combination of several regression trees. They predict numbers at leaf nodes using several non-linear regression. While Support Vector Regression (SVR) uses the principle as SVM for classification by generalization bounds given for regression. A limit of tolerance (epsilon) is supposed in the prediction of real numbers while a loss function is determined that avoid errors which are stimulated within a specific distance to the true value.

The results of the regression analysis for human age estimation using RF and SVR is given in Table 3. Thus, RF regression shows the best efficiency with the smallest root mean square of error (RMSE) value 2.6 years for S-BS for the dataset and the variance which corresponds to 4.14 years. SVR has a lower accuracy than RF for S-BS for $t_1 = 20s$ protocol in our case. Therefore, the regression model built using RF in 10-fold cross-validation is the most efficient approach for age estimation using the auditory perception.

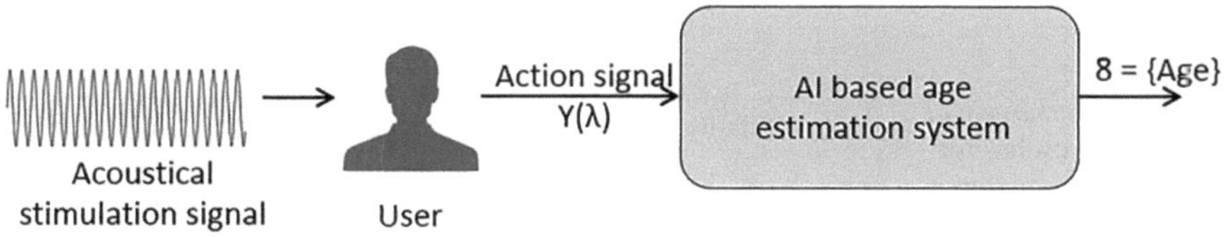

Fig. 3 Auditory perception based human estimation system

Table 3 Performance of age estimation

Dataset	Regression forest		Support vector machine	
	RMSE		Variance	
S-BS(20 s)	2.6	6.6	4.1	27.1
S-BS(40 s)	2.8	6.7	4.3	25.9
S-BS(60 s)	3.8	5.8	5.1	15.7
H-BS	3.1	5.5	4.4	17.7
H-RS	3.8	6.9	8.7	28.8
H-LS	3.2	5.8	4.9	19.4

4.4 Comparison of Existing Biometric Modalities

In Tables 4 and 5 comparison have been presented of the existing biometric modalities.

For human age classification, the existing modalities of biometrics are shown in Table 4. The classification performance using face traites shows 66.6% accuracy for seven age groups [2]. While voice trait shows 70% of accuracy for five age groups and 27–54% for seven age groups [1, 2]. Keystroke dynamic shows the highest accuracy

Table 4 Comparison of classification performances of different biometric modalities

Modality	References	Two age groups	Three age groups	Four age groups	Five age groups	Seven age groups
Face	[2]					66.6%
Voice	[1, 2]				70%	27–54%
Keystroke	[11, 12]	72% 90%		64%		
Auditory perception			92%		86%	

Table 5 Comparison of existing approaches for human age estimation

Modality	References	Mean absolute error
Face	[3]	4.87
		5.82
		3.64
		5
Voice	[1]	4.7–10
Gait	[9]	6
	[10]	8.2
DNA	[4]	4
Auditory perception		2.6

among all the previous existing approaches with an accuracy of 72–90% for two age groups and 64% for four age groups.

As shown in Table 5, using auditory perception based approach has a smaller mean absolute error for human age estimation than the other existing approaches based on face [3], voice [1], gait [9, 10] and DNA [4] is obtained. While the auditory perception based human age estimation approach shows the best results age classification and estimation than existing biometric approaches.

5 Verification/Identification System

Several applications are possible to the proposed approach of age estimation based on the auditory perception, one of these applications concern identity verification. A control access web filter is proposed which is known as BiometricAccessFilter.

In Fig. 4, the pipeline of the BiometricAccessFilter system is explained, how a user creates an account on the online web interface. It needs, some information to be inserted manually like name, and password. After creating an account successfully, on the Login page age verification system is designed to give the right to the user to access the webpage if the user has the right age. If the estimated age of the user is higher than the standard threshold of age (years) then the access is denied and if the user's age is lower than the standard threshold, the account will be created for further use.

BiometricAccessFilter can restrain the access of adults to the virtual environment of kids and on the other hand, it prevents child sexual abuse in the world of internet. With the help of this web application, a ground is created to change the approach and new ways of developing more secure web applications are presented using smart biometric techniques. While it can help in a wide range of applications in detection of cybercrime. It can reduce the loopholes of traditional identification systems.

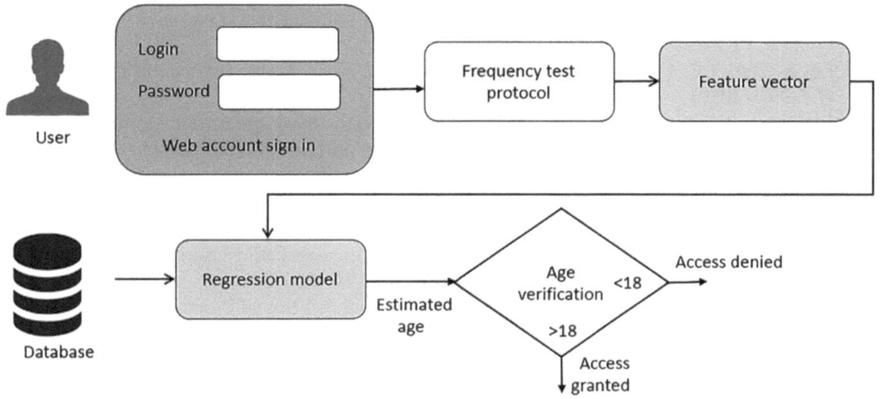

Fig. 4 BiometricAcessFilter application

6 Conclusion

A new hidden biometrics study is introduced corresponding to biometric age classification and estimation based on the auditory perception. Several protocols have been tested. The best protocol correspond to bilateral stimulation with a duration of 20 s. A good classification rate of 92 and 86% is attained for three and five age groups. A robust regression model is also built a root mean square of error of 2.6 years. Considering the results for age estimation, a biometricaccessfilter is designed which is entirely a new approach to access control systems in the field of biometrics. BiometricAccessControl system is based on auditory perception and it can limit the access of user within the specific age group. Furthermore, this proposal is sensitive to attacks, a user can easily fool the system about their age.

References

1. Moyse, E.: Age estimation from faces and voices: a review. Psychol. Belgica **54**(3) (2014)
2. Eidinger, E., Enbar, R., Hassner, T.: Age and gender estimation of unfiltered faces. IEEE Trans. Inf. Forensics Secur. **9**(12), 2170–2179 (2014)
3. Lanitis, A., Draganova, C., Christodoulou, C.: Comparing different classifiers for automatic age estimation. IEEE Trans. Syst. Man Cybern. Part B (Cybern.) **34**(1), 621–628 (2004)
4. Freire-Aradas, A., Phillips, C., Lareu, M.V.: Forensic individual age estimation with DNA: from initial approaches to methylation tests. Forensic Sci. Rev. **29**(2) (2017)
5. Williams, G.: A review of the most commonly used dental age estimation techniques. J. Forensic Odontostomatol. **19**(1), 9–17 (2001)
6. Shafran, I., Riley, M., Mohri, M.: Voice signatures. In: 2003 IEEE Workshop on Automatic Speech Recognition and Understanding, 2003. ASRU'03, pp. 31–36. IEEE (2003)
7. Metze, F., Ajmera, J., Englert, R., Bub, U., Burkhardt, F., Stegmann, J., Muller, C., Huber, R., Andrassy, B., Bauer, J.G., Littel, B.: Comparison of four approaches to age and gender recognition for telephone applications. In: IEEE International Conference on Acoustics, Speech and Signal Processing, 2007. ICASSP 2007, vol. 4, pp. IV-108. IEEE (2007)
8. Dobry, G., Hecht, R.M., Avigal, M., Zigel, Y.: Supervector dimension reduction for efficient speaker age estimation based on the acoustic speech signal. IEEE Trans. Audio Speech Lang. Process. **19**(7), 1975–1985 (2011)
9. Lu, J., Tan, Y.P.: Gait-based human age estimation. IEEE Trans. Inf. Forensics Secur. **5**(4), 761–770 (2010)
10. Makihara, Y., Okumura, M., Iwama, H., Yagi, Y.: Gait-based age estimation using a whole-generation gait database. In: 2011 International Joint Conference on Biometrics (IJCB), pp. 1–6. IEEE (2011)
11. Tsimperidis, G., Katos, V., Rostami, S.: Age detection through keystroke dynamics from user authentication failures. Int. J. Dig. Crime Forensics (IJDCF) **9**(1), 1–16 (2017)
12. Uzun, Y., Bicakci, K., Uzunay, Y.: Could We Distinguish Child Users from Adults Using Keystroke Dynamics? (2015). arXiv preprint arXiv:1511.05672
13. Smith, S.W.: The Scientist and Engineer's Guide to Digital Signal Processing, p. 35 (1997)
14. Zwicker, E.: Subdivision of the audible frequency range into critical bands (frequenzgruppen). J. Acoust. Soc. Am. **33**, 248 (1961)
15. Stuart, R., Howell, P.: Signals and Systems for Speech and Hearing. 2nd edn., pp. 163. BRILL (2011)
16. Rossing, T.: Springer Handbook of Acoustics, 1st edn., pp. 747–748. Springer (2007)

17. Ilyas, M., Othmani, A., Nait-Ali, A.: Human age estimation using auditory system through dynamic frequency sound. In: IEEE 2nd International Conference on Bio-engineering for Smart Technologies (BioSMART) (2017)
18. Stockwell, C.W., Ades, H.W., Engström, H.: XCVII patterns of hair cell damage after intense auditory stimulation. Ann. Otol. Rhinol. Laryngol. Suppl. **78**, 1144–1168 (2017)
19. Manley, G.A., van Dijk, P.: Frequency selectivity of the human cochlea: suppression tuning of spontaneous otoacoustic emissions. Hear Res. **336**, 53–62 (2016)
20. Paolis, A.D., Bikson, M., Nelson, J.T., de Ru, J.A., Packe, M., Cardoso, L.: Analytical and numerical modeling of the hearing system: Advances towards the assessment of hearing damage. Hear. Res. **349**, 111–128 (2017)
21. Barbosa de Sá, L.C., Lima, M.A.M.T., Tomita, S., Frota, S.M.M.C., Santos, G.A., Garcia, T.R.: Analysis of high frequency auditory thresholds in individuals aged between 18 and 29 years with no ontological complaints. Rev. Bras. Otorrinolaringol. **73**, 2 (2007)
22. Breiman, L.: Random forests. Mach. Learn. **45**, 123–140 (2011)
23. Guyon, I., Saffari, A., Dror, G., Cawley, G.: Model selection: beyond the bayesian–frequentist divide. JMLR **11**, 61–87 (2010)
24. Anguita, D., Ghio, A., Ridella, S., Sterpi, D.: K-fold cross validation for error rate estimate in support vector machines. In: Proceedings of the International Conference on Data Mining, USA, pp. 291–297 (2009)
25. Dietterich, T.G.: Approximate statistical tests for comparing supervised classification learning algorithms. Neural Comput. **10**(7), 1895–1923 (1998)
26. Statnikov, A., Tsamardinos, I., Dosbayev, Y., Aliferis, C.F.: GEMS: a system for automated cancer diagnosis and biomarker discovery from microarray gene expression data. Int. J. Med. Inform **74**, 491–503 (2005)
27. Scheffer, T.: Error estimation and model selection. Ph.D. Thesis, Technischen Universität Berlin, School of Computer Science (1999)

Multi-, Hyper-Spectral Biometrics Modalities

Mohsen Ardabilian, Abdel-Malek Zine and Shiwei Li

Abstract In this chapter, it will be introduced the different categories of multi-hyper-spectral imaging approaches for biometric modalities. Afterwards, indirect approach will be considered, namely: prerequisites concepts on physics and computer graphics, physical theory for the light-skin interaction models and finally the related applications of multi-hyper-spectral imaging.

1 Introduction

The multi-spectral modalities in biometrics refer to approaches that use other complementary spectral bands than those used in conventional RGB sensors. The motivation is to reach discriminant characteristics, not visible to naked eye and necessarily to the conventional RGB cameras. An extension is the hyper-spectral imaging that would ensure higher performance and accuracy through an increase in the spectral bands obtained by the sensor and analyzed thereafter. In the both cases, each pixel of the image contains spectral information. Thus, a third dimension of values is added to the two-dimensional spatial image, generating a three-dimensional data cube, also referred to as hypercube data or as an image cube. The common RGB color image can be considered as a simple example of a three-dimensional data cube. Usually, the data cubes contain reflectance, absorption, or fluorescence information. To achieve such a data cube more sophisticated sensors are required whose cost is often higher. But, this obstacle to the development of more efficient but also safer biometric systems dissipates with the progress in sensors.

The acquisition stage is then followed by the analysis one. Multi or hyper-spectral analysis approaches can be split into two main categories, direct and indirect approaches. Direct approaches rely on the application of a known pattern recognition method, i.e. feature extraction and selection followed by classification step, on

---

M. Ardabilian (✉) · S. Li
LIRIS Laboratory—UMR 5205 CNRS, Université de Lyon, Ecole Centrale de Lyon, Lyon, France
e-mail: Mohsen.Ardabilian@ec-lyon.fr

A.-M. Zine
ICJ—UMR 5208 CNRS, Université de Lyon, Ecole Centrale de Lyon, Lyon, France

A. Nait-ali (ed.), *Hidden Biometrics*, Series in BioEngineering,
https://doi.org/10.1007/978-981-13-0956-4_8

acquired multi- or hyper-spectral data of an individual. An example of direct approach is the vein pattern recognition. Indirect approaches, on the other hand, consist of estimating of the physical or biological properties of the tissues. These properties are often the parameters of a model of interaction between a source of energy and the tissues. An example is the physical model of light interaction with skin tissues. The set of parameters that, by using an interaction model, allows to reproduce an observation is retained as the estimated properties of the observed tissues.

In the next sections we first introduce the different categories of multi-hyper-spectral imaging approaches. Then we focus on an indirect approach: beginning with prerequisites concepts on physics and computer graphics, we continue on physical theory for the light-skin interaction models and the related applications of multi-hyper-spectral imaging including three categories of algorithms implementing light-skin interaction model. Finally, we conclude the chapter with related applications and discussions.

2 Multi-Hyper-Spectral Devices

A multitude technical approaches for Multi-Hyper-Spectral Imaging (MHSI) have been proposed [1–3]. These approaches can be categorized into multiple distinct classes based on which axes of the data-cube are sampled at a given instant of time. For example, to build up a full data-cube, dispersive approaches instantaneously sample along the spectral axis and along one spatial axis, while the remaining spatial axis is scanned in time.

More precisely, in a line scanning the light passing through a slit is dispersed by a grating or prism before being imaged on a two-dimensional detector array. To cover a scene, the line scanning process is sequentially displaced to cover and produce the full data cube. Using wavelength-tunable filters, a first category of hyper-spectral sensors acquires instantaneously the 2D spatial information within the specified wavelength. For instance, wavelength-tunable filter can be designed based on liquid crystals [4, 5], holographic polymer-dispersed liquid crystals [6, 7], Fabry-Perot interferometer [1], acousto-optical tunable filter (AOTF), or a motorized filter wheel.

"Snapshot" filters attempt to sample the full data-cube instantaneously [2]. For a given image sensor area, a fixed tradeoff between spatial and spectral resolution must be chosen. Cameras based on tiled or mosaic filter arrays are simply extensions of the common RGB Bayer mosaic, and are already commercially available.

According to Spectral imaging based on Fourier transform spectroscopy, the entire spectrum is sampled at different points of the Fourier space, producing an interferogram that is Fourier transformed to give the optical spectrum. In general, this is done using Michelson interferometry. To suit the application without compromising optical throughput, the Spatial/Spectral resolution power and imaging speed are traded off by software.

Compressive spectral imaging [8, 9], for instance through deconvolution of overlapped spatial/spectral information [10], was also proposed for spectral multiplexing.

High signal to noise is used to overcome the ill-conditioned problem of deconvolution.

In the following, we propose two approaches and HSI devices that could potentially meet needs and requirements in biometrics. Indeed, these two devices are low cost, operate in snapshot mode, are miniaturized or can be easily, possess sufficient spatial resolution and a short acquisition time.

The first device is proposed by researchers from Vienna University of Technology and University of Arizona [11]. They showed that by relaxing the requirement of sub-second exposure times, and instead using standard High Dynamic Range (HDR) image acquisition, one can perform CTIS spectral imaging using unmodified consumer cameras. The other principal advantages of CTIS are maintained, i.e., avoiding any opto mechanical components, requiring only a simple optical path, and having all information recorded in one HDR image. In addition, other advantages of this solution are its low cost and compact design. These features allow a fully portable camera objective that can be cheaply constructed from off-the-shelf photographic components, leading to drastically lower cost than any other spectral imaging device while every component can still be used for its original purposes. Despite the severe compromises made, the proposed solution offers a spectral resolution of 0.8 nm in slit configuration, i.e. masking out all but a single pixel-wide column, and 4.89 nm resolution (54 spectral bands) in a hyperspectral configuration of 120 × 120 pixels. These resolutions are competitive with non-imaging spectrometers as well as dedicated CTIS devices, but at the expense of exposure times in the range of a few seconds rather than milliseconds, depending on the dynamic range of the scene. Such resolutions cannot be achieved with filter-based systems.

The second HSI device [12], proposed by PARC, takes advantage of the tunable path delay between orthogonal polarization states of a liquid crystal variable retarder. An interferogram is generated simultaneously at each pixel based on this varied path delay between the polarization states. A data set consisting of a series of images is recorded while varying the path delay; Fourier transforming the data set with respect to the path delay yields the hyper spectral data-cube. This concept is demonstrated with a prototype imager consisting of a liquid crystal variable retarder integrated into a commercial 640 × 480 pixel CMOS camera. The prototype can acquire a full hyper-spectral data-cube in 0.4 s, and is sensitive to light over a 400–1100 nm range with a dispersion dependent spectral resolution of 450–660 cm^{-1}. Similar to Fourier transform spectroscopy, the imager is spatially and spectrally multiplexed and therefore achieves high optical throughput. Additionally, the common path nature of the polarization interferometer yields a vibration-insensitive device. This concept allows for the spectral resolution, imaging speed, and spatial resolution to be traded off in software to optimally address a given application. The simplicity, compactness, potential low cost, and software adaptability of the device may enable a disruptive class of hyper-spectral imaging systems.

3 Light-Skin Interaction Based Skin Analysis

Skin is a complex organ but also the largest part seen first by others. Its appearance is affected by many factors known as intrinsic (or genetic, endogenous) and extrinsic (or environmental, exogenous) factors; from an individual's appearance, one can generally glean a large amount of information such as gender, ethnicity, age, health condition, etc. It has been intensively studied in life, physical and computer sciences, biometrics, because some medical diagnosis and treatments of dermatological diseases are based on the measurement of the substance concentration in skin. In cosmetic industry, some skin-related process simulations are required in product development and evaluation. In computer science, skin appearance studies are also very important mainly in realistic images and video synthesis. In biometrics, skin texture and other behaviors are used for identification, authentication, countermeasure. In medicine, skin characterization and analysis are used for the diseases diagnosis and surgery guidance.

Among all the extrinsic stimulus which produce observable responses, visible light is a perfect choice for studying and modelling through light-skin interactions. Because visible light is easy to produce for controlled conditions and has sufficient optical penetration properties to provide enough subcutaneous information. Analytic light-skin interaction models in visible domain are widely studied in computer graphics, and are combined with physical or biological theories for realistic images or videos synthesis purposes. The light-skin interaction models can be roughly classified as believable or predictable [13]. Believable models are biophysically inspired simulations, they aim to produce aesthetically realistic and plausible results; the predictable models are derived from biophysical-based simulations, whose validity and accuracy are evaluated through comparisons with measured data, from which one can also extract skin biological properties, and can gain insight into the light-skin interaction phenomena. Although predictable models is close to reality and is expressed by biological parameters, that is not to say the predictable models overwhelm believable models. Because errors may be introduced by approximations, or by other processes during the simulation process. Sometimes believable models can yield more satisfying results if the parameters are correctly set.

In the next section, we first provide physical background for light-skin interaction modelling, including some basic concepts of optics and radiative transfer theory; then the presentation is followed by a general introduction of several light-skin interaction models.

3.1 Basic Optical Concepts and Radiative Transfer Theory

Light (refers to visible light in this chapter) is electromagnetic radiation, a form of energy, which is distinguished by wavelength ranging from 400 to 700 nm. When a beam of light hits the medium interface, as shown in Fig. 1, reflection and transmission

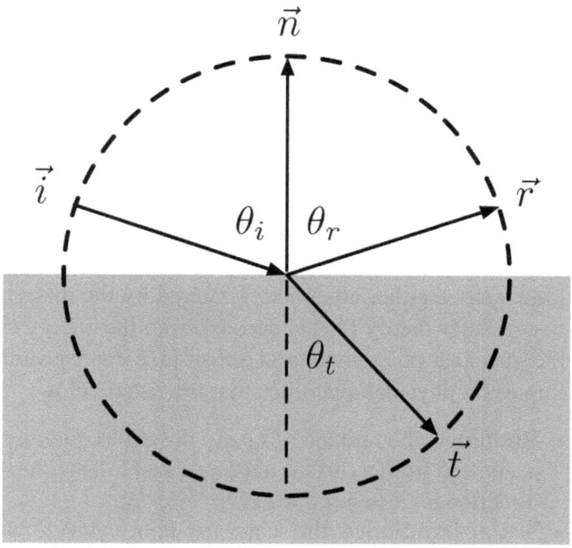

Fig. 1 Geometric parameters for light-material interaction at interface

(or refraction), occurs, which is described by the Fresnel equations [1]. The Fresnel diffuse reflectance can be separated into a *s*-polarised term and a *p*-polarised term. The reflectance for *s*-polarised light is

$$R_s = \left| \frac{n_1 \cos\cos\theta_i - n_2 \cos\cos\theta_t}{n_1 \cos\cos\theta_i + n_2 \cos\cos\theta_t} \right|^2, \tag{1}$$

while the reflectance for *p*-polarised light is

$$R_p = \left| \frac{n_1 \cos\cos\theta_t - n_2 \cos\cos\theta_i}{n_1 \cos\cos\theta_t + n_2 \cos\cos\theta_i} \right|^2. \tag{2}$$

If the incident is unpolarised, the reflectance will be

$$R = \frac{R_s + R_p}{2}. \tag{3}$$

where n_i and n_t are the refractive index of the two related mediums, θ_i represents the incident angle, and θ_t is the transmission angle which can be determined by the Snell's law:

$$n_i \sin\ \sin\theta_i = n_t \sin\ \sin\theta_t. \tag{4}$$

As a consequence of the energy conservation law, the transmittance is given by

$$T = 1 - R. \tag{5}$$

Particularly, the incident angle is called the "critical angle θ_c", if the transmission angle becomes 90°, given by

$$\theta_c = \arcsin \arcsin\left(\frac{n_t}{n_i}\right). \quad (6)$$

In the case that incident angle is greater than θ_c, all incident light is reflected back to the incident medium, known as internal reflection or total reflection.

Once the light penetrates into a medium, it interacts with particles of the material, light will be either absorbed, scattered by the material or propagate further. Radiative transfer theory [14] is derived from the energy conservation law, and describes this physical phenomenon of subsurface energy variations. This theory has several important physical quantities that are listed below:

- **Radiance**: the radiance $L(x, \vec{\omega})$ describes quantity of radiation passes through or is emitted from a surface at point x and falls within a given solid angle in specified direction $\vec{\omega}$, measured in (W m^{-2} sr^{-1});
- **Scalar irradiance**: the scalar irradiance $\phi(x, \vec{\omega})$ is the power of electromagnetic radiation per unit area incident on a surface, $\phi(x) = \int_{4\pi} L(x, \vec{\omega}) d\Omega$ is measured in (W m^{-2});
- **Vector irradiance**: the vector irradiance $\vec{E}(x)$ represents the irradiance pointing in the prevalent direction of the energy flow, $\vec{E}(x) = \int_{4\pi} L(x, \vec{\omega})\vec{\omega} d\Omega$ is measured also in (W m^{-2});
- **Phase function and anisotropy factor**: the phase function $p(x, \vec{\omega}_i, \vec{\omega}_0)$ signifies the probability of light with propagation direction $\vec{\omega}_i$ being scattered into solid angle $d\Omega$ around $\vec{\omega}_0$, and depends usually only on the angle between the incident angle ω_i and scattered angle ω_0; the anisotropy factor g can be expressed as $g = \int_{4\pi} (\vec{\omega}_i \cdot \vec{\omega}_0) p(\vec{\omega}_i \cdot \vec{\omega}_0) d\Omega$.

In the definitions above, the integral over 4π signifies the surface integral of the unit sphere. During the light propagation, the radiative transfer equation simply says that as the light travels, it loses energy to absorption, gains energy by emission, and redistributes energy by scattering (in-scattering and out-scattering), see Fig. 2. The variation of radiance is the sum of these four cases, by dividing $dS(x)$ on both sides, we arrive at the radiative transfer equation in physics, also refers to the volume rendering equation in computer graphics. In the case of homogenous medium, the equation can be expressed as

$$\left(\vec{\omega} \cdot \vec{\nabla}\right) L(x, \vec{\omega}) = -\sigma_t L(x, \vec{\omega}) + \sigma_s \int_{4\pi} p\left(\vec{\omega}' \cdot \vec{\omega}\right) L\left(x, \vec{\omega}'\right) d\Omega' + Q(x, \vec{\omega}). \quad (7)$$

where σ_a and σ_s are the absorption and scattering coefficients which describe the medium optical properties. The extinction coefficient σ_t is defined as, $\sigma_t = \sigma_a + \sigma_s$. $Q(x, \vec{\omega})$ is the volume source distribution.

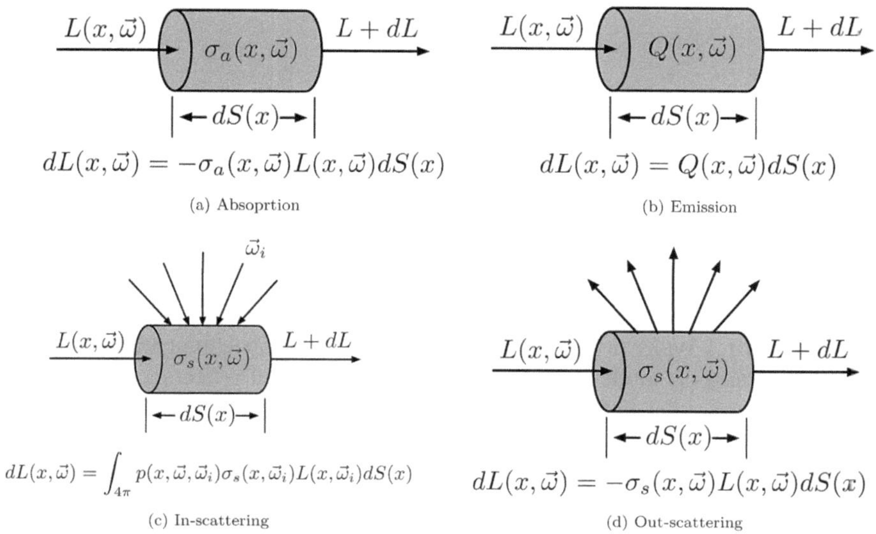

Fig. 2 Radiative transfer theory

To gain insight into the volumetric behavior of light propagation, the integral overall directions $\vec{\omega}$ at point x yields

$$\vec{\nabla}.\vec{E}(x) = -\sigma_a \phi(x) + Q_0(x). \tag{8}$$

where $Q_0(x) = \int_{4\pi} Q(x, \vec{\omega})d\Omega$ is called the 0th-order source distribution. If the medium is a finite or semi-infinite region, the irradiant exiting the medium at point x_s is equal to $\vec{n} \cdot \vec{E}(x_s)$ [15]. Light that penetrates into the medium, will leave the medium by reflection and transmission, or be absorbed by the medium. The energy conservation indicates that the incident light power Φ_i is equal to the sum of body reflection Φ_r, body transmission Φ_t and body absorption Φ_a power and specular reflection Φ_s.

$$\Phi_i = \Phi_s + \Phi_r + \Phi_t + \Phi_a. \tag{9}$$

3.2 Light-Skin Interaction Models

In computer graphics, although different light-skin interaction models are derived from the energy conservation law and transfer theory of radiation; different assumptions or approximations, and different forms of energy conservation law (Eq. 9) lead to various types of light-skin interaction models.

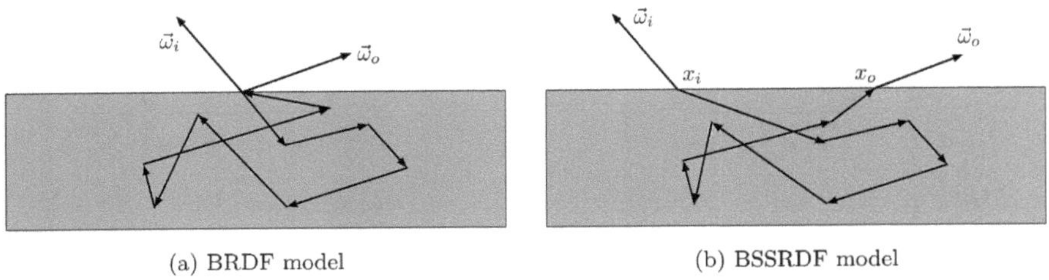

Fig. 3 BRDF model

There exist 2 types of models describing local or global light transports. The local light transport models (refers to BRDF, bidirectional reflectance distribution function) assume that light enters and leaves the medium at the same location as entering, after interaction with the medium, see Fig. 3a. BRDF can be used to describe how light is reflected at an opaque surface, it takes the negative incoming light direction $\vec{\omega}_i$, and a outgoing direction $\vec{\omega}_0$ as input; returns the ratio of reflected radiance $L_r(\vec{\omega}_0)$ exiting along $\vec{\omega}_0$ to the irradiance incident $\phi_i(\vec{\omega}_i)$ on the surface along $\vec{\omega}_i$. The BRDF f_r is given by

$$f_r(\vec{\omega}_i, \vec{\omega}_0) = f_r(\theta_i, \varphi_i; \theta_0, \varphi_0) = \frac{dL_r(\vec{\omega}_r)}{d\phi_r(\vec{\omega}_i)} = \frac{dL_r(\vec{\omega}_0)}{dL_i(\vec{\omega}_i)\cos\cos\theta_i d\Omega_i}. \quad (10)$$

The global light transport models (refers to BSSRDF, bidirectional subsurface scattering reflectance distribution function) assume that light enters and leaves the medium at different locations, see Fig. 3b. BSSRDF can be used to describe how light transports in the subsurface of a translucent medium, it takes the entering point x_i, negative incident direction $\vec{\omega}_i$, outgoint point x_0 and the outgoing direction $\vec{\omega}$ as input; returns the ratio of reflected radiance $L_0(x_0, \vec{\omega}_0)$ at point x_0 in direction $\vec{\omega}_0$ to the incident irradiance $\phi_i(x_i, \vec{\omega}_i)$ at point x_i from direction $\vec{\omega}_i$. Generally, the BRDF can be considered as a special case (for opaque materials) or an approximation (for translucent materials) of BSSRDF.

Specular Reflection Model is the simplest reflectance model that all incident light energy is reflected as specular reflection in a single direction, and is derived from the geometric optics. The reflected light direction $\vec{r}$ and negative light direction $\vec{l}$ are symmetric to the surface normal $\vec{n}$, see Fig. 4. As there will be reflected energy only if the negative view direction $\vec{v}$ is equal to the reflected direction $\vec{r}$, the specular BRDF is simply a double-delta function

$$f_r(\theta_i, \varphi_i; \theta_0, \varphi_0) = \alpha_{sp}\delta(\theta_i - \theta_0)\delta(\varphi_i + \pi - \phi_0). \quad (11)$$

where α_{sp} is called the specular albedo and the reflected direction can be calculated directly by

$$\vec{r} = 2\left(\vec{l} \cdot \vec{n}\right)\vec{n} - \vec{l}. \quad (12)$$

Fig. 4 Specular reflection

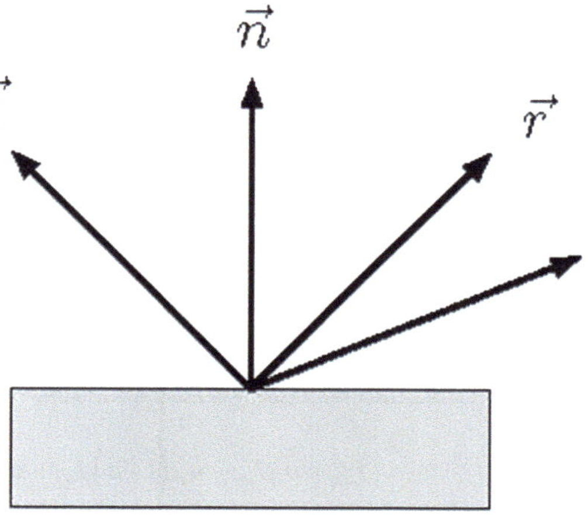

Lambertian Model is the simplest diffuse reflectance model which is first introduced in J. H. Lambert's book [16]. He defines an ideal diffuse reflecting surface (only diffuse reflection) called Lambertian surface (perfect diffuse reflector), which either emits or reflects radiation into all directions of a hemisphere with constant radiance obeying Lambertian's cosine law shown in Fig. 5.

$$\vec{l} \cdot \vec{n} = \cos\cos 0^\circ \quad \vec{l} \cdot \vec{n} = \cos\cos 30^\circ \quad \vec{l} \cdot \vec{n} = \cos\cos 45^\circ$$

The Lambertian BRDF is simply a constant.

$$f_r(\theta_i, \varphi_i; \theta_o, \varphi_o) = \frac{\alpha_d}{\pi}. \tag{13}$$

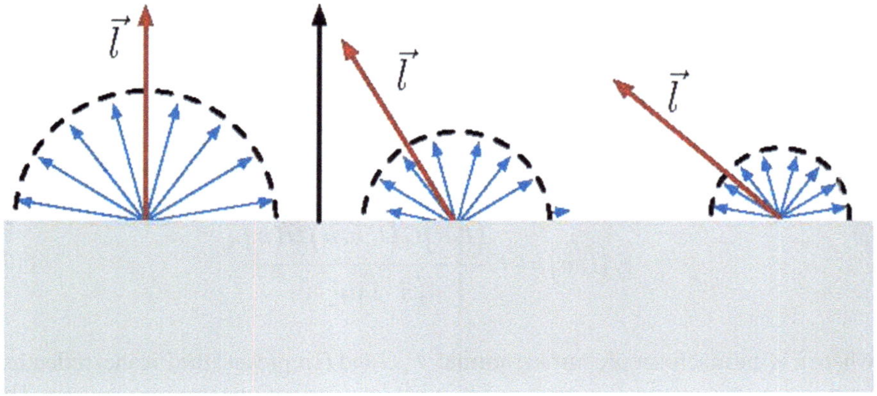

Fig. 5 Lambertian cosine law

The diffuse light intensity depends on the angle between the negative incident direction $\vec{l}$ and surface orientation $\vec{n}$

$$I_{\vec{l}} = \alpha_d I_{\vec{l}} \left(\vec{n} \cdot \vec{l} \right). \tag{14}$$

where α_d is the diffuse constant determining the percent of reflected light. Note that the Lambertian BRDF does not take specular reflection into account, so it is usually combined with specular reflection model.

Phong Model and Blinn-Phong Model [17] is an empirical local transport model. It separates the source and reflected light energy into 3 parts: ambient light, diffuse light and specular light. The intensity of a surface point I_p can be computed by

$$I_p = \alpha_a I_a + \alpha_d \left(\vec{l} \cdot \vec{n} \right) I_d + \alpha_s (\vec{r} \cdot \vec{v})^m I_s. \tag{15}$$

where α_a, α_d and α_s represent albedo for ambient, diffuse and specular reflection; I_a, I_d and I_s are the ambient, diffuse and specular light intensity respectively. m denotes the shininess of the surface, very large m leads to mirror-like reflection. Vector symbols are the same with that of specular reflection model, same as following in this section. The term $(\vec{r} \cdot \vec{v})^m$ continuously changes when the light source or view point moves, this value needs to be re-computed for every pixel. In most cases, the relative position between light and viewer remains the same, or the viewer and light source are at infinity; J. F. Blinn made a modification [2] by substituting the computational heavy term $(\vec{r} \cdot \vec{v})^m$ by $\left(\vec{n} \cdot \vec{h} \right)^{m'}$, where $\vec{h} = \frac{\vec{l}+\vec{v}}{\|\vec{l}+\vec{v}\|}$ is called the half vector; and the shininess factor m by a larger value m'. This significantly reduces the computation in the rendering process, and can be computed only once and used for the entire frame. The Blinn-Phong BRDF is given by

$$f_r = \alpha \frac{m+2}{2\pi} \delta. \tag{16}$$

where α denotes the albedo and $(m2)/2\pi$ is the energy normalisation term, which assures that 100% of the incident energy is reflected for $\alpha = 1$.

Torrance-Sparrow Model [18, 19] is a BRDF especially for rough surfaces. It models the medium surface as symmetric, longitudinal, infinitely-long V-grooves at a microscopic level, see Fig. 6. The Torrance-Sparrow BRDF is given by

$$f_r\left(\vec{l}, \vec{v}\right) = F \frac{\left(\vec{l}, \vec{h}\right) G\left(\vec{l}, \vec{v}, \vec{h}\right) D\left(\vec{h}\right)}{4\left(\vec{n} \cdot \vec{l}\right)(\vec{n} \cdot \vec{v})}. \tag{17}$$

where $\vec{n}$ is the macroscopic surface normal; F, G and D represent the Fresnel reflection term, geometric attenuation term and microfacet distribution term respectively. The

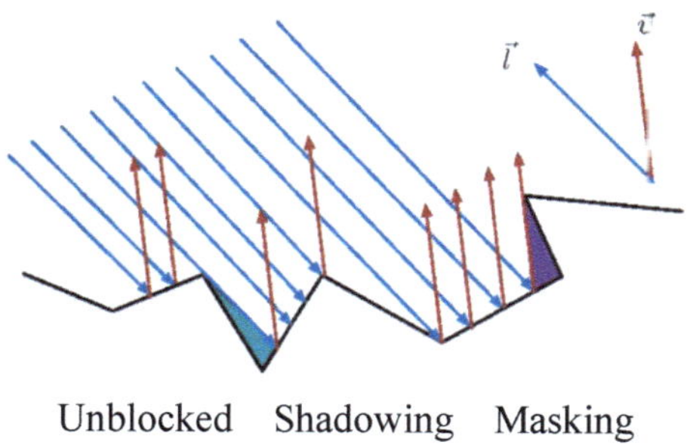

Fig. 6 Torrance-sparrow model

geometric attenuation term describes how much light that escapes from the surface, and is computed for each case in Fig. 6:

- Unblock: no interference, all light will be reflected, $G_u = 1$;
- Shadowing: some of the incident light is masked off, $G_s = \frac{2(\vec{n}\cdot\vec{h})(\vec{n}\cdot\vec{l})}{\vec{v}\cdot\vec{h}}$;
- Masking: some of the reflected light is intercepted, $G_m = \frac{2(\vec{n}\cdot\vec{h})(\vec{n}\cdot\vec{v})}{\vec{v}\cdot\vec{h}}$.

The minimal value is chosen as the final geometric attenuation term $G = \min(G_u, G_s, G_m)$.

The micro facet distribution term $D\left(\vec{h}\right)$ gives the probability density of microfacet oriented along the direction $\vec{h}$. There exist several distributions, but the Beckmann distribution function is widely used in skin appearance modelling in computer graphics, the Beckmann distribution D_B is given by:

$$D_B\left(\vec{h}\right) = \frac{e^{\frac{\left(\vec{n}\cdot\vec{h}\right)^2-1}{m^2\left(\vec{n}\cdot\vec{h}\right)^2}}}{\pi m^2\left(\vec{n}\cdot\vec{h}\right)^4}. \tag{18}$$

where m is the root-mean-square slope of all microfacets, representing the roughness of the surface.

Kubelka-Munk Model [20] was first developed for paint. It shows a simple relationship between the optical properties of paint (absorption coefficient σ_a and scattering coefficient σ_s) and its overall reflectance. This model assumes that the scattering and absorbing media are uniformly distributed, and describes the subsurface light transport on the basis of two oppositely directed fluxes, as shown in Fig. 7. Consider light of intensity I_i incident on a slab of thickness d. At a distance x from the top surface of the medium, there is a lamina of thickness dx. The scattered light is incident on it, which is travelling both upwards and downwards through it with intensities i_r and i_t, respectively. The effect of the medium in a thin element d_x on i_r and i_t is to:

- decrease i_t by $(\sigma_a + \sigma_s)i_t dx$ due to absorption and scattering;
- decrease i_r by $(\sigma_a + \sigma_s)i_r dx$ due to absorption and scattering;

Fig. 7 Kubelka-Munk Model

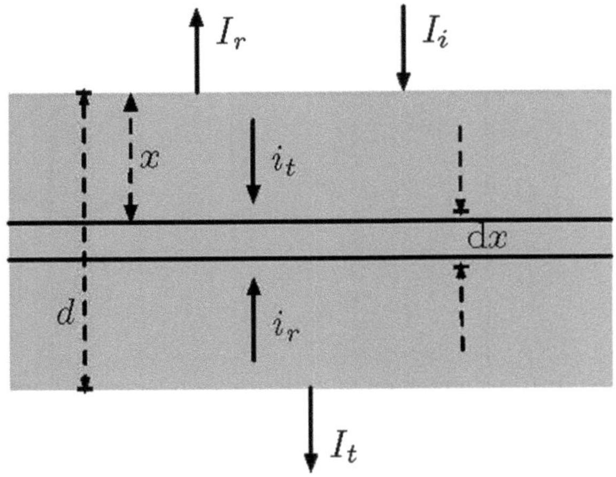

- increase i_t by $\sigma_s i_r dx$ due to scattering from i_r;
- increase i_r by $\sigma_s i_t dx$ due to scattering from i_t.

So we have:

$$di_t = -(\sigma_a + \sigma_s)i_t dx + \sigma_s i_r dx, -di_r = -(\sigma_a + \sigma_s)i_r dx + \sigma_s i_t dx. \quad (19)$$

The overall reflectance and transmittance can be expressed by exponential functions:

$$R(d) = \frac{(1-\beta^2)(e^{Kd} - e^{-Kd})}{(1+\beta)^2 e^{Kd} - (1-\beta)^2 e^{-Kd}}, T(d) = \frac{4\beta}{(1+\beta)^2 e^{Kd} - (1-\beta)^2 e^{-Kd}}. \quad (20)$$

where $K = \sqrt{4\sigma_a^2 + 8\sigma_a\sigma_s}$ and $\beta = \sqrt{\frac{2\sigma a}{2\sigma_a + 4\sigma_s}}$. Although Kubelka-Munk models the subsurface light transport, it assumes that there is no lateral scattering, it is still a BRDF.

Monte Carlo method is a statistical approach to the solution of integral-differential equations that occur in various branches of natural sciences. Modelling light propagation with Monte Carlo is a flexible yet rigorous approach [21, 22]. The main idea of Monte Carlo method is called importance sampling [8]. If we need to evaluate the integral

$$I = \int_a^b f(x)dx. \quad (21)$$

But sometime $f(x)$ is not analytic, in this case, we can rewrite the integral as

$$I = \int_a^b \frac{f(x)}{P(x)} P(x) dx. \tag{22}$$

where $P(x)$ represents the importance function, and needs to satisfy conditions:

- $P(x) \geq 0$ for each $x \in [a, b]$, for which $q(x) \neq 0$;
- $\int_a^b P(x)dx = 1$;
- $\frac{f(x)}{P(x)} < \infty$ almost everywhere.

A technique called warping [23] is often used to solve this problem. It consists of generating uniform distributed random samples, denoted by ξ, in a canonical space [0, 1]. These samples are then transformed so that their distribution matches P. Assuming that a multi-layered material with optical properties: absorption σ_a^i and scattering coefficients σ_s^i, anisotropy factor g_i, and refractive index n for i-th layer.

A scheme for modelling photon flow in a medium with Monte Carlo method is shown in Fig. 8.

In this simulation, random sampling is required in 3 steps: determining step size, determining reflection or transmission, and scattering.

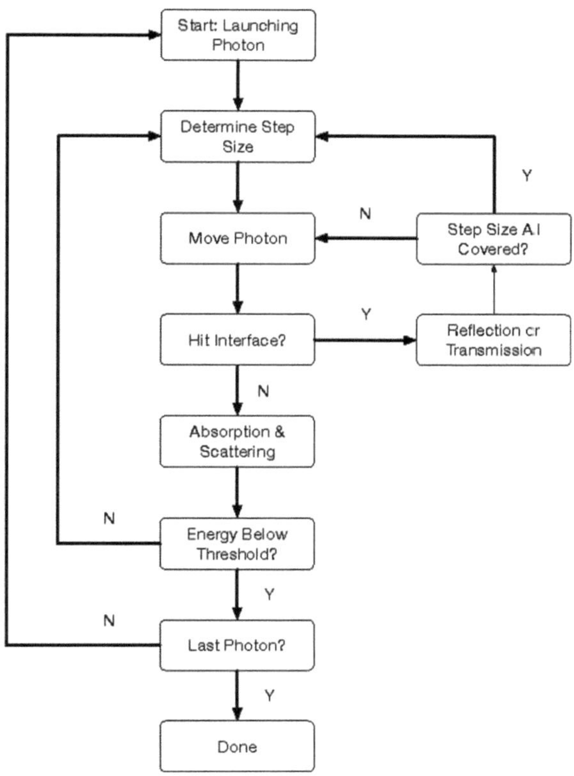

Fig. 8 Monte Carlo simulation of light transport in multi-layered material

- Determining step size: the step size of the photon for one run is determined by $s = \frac{-\ln \ln \xi}{\sigma_t}$, where $\sigma_t = \sigma_a + \sigma_s$ is called the extinction coefficient;
- Determining reflection or transmission: if the generated random number ξ is smaller than the internal reflectance R_{in}, the photon is internally reflected; otherwise it will transmit into the next medium;
- Scattering: the scattering phase function of i-th layer is described by Henyey-Greenstein function, see Fig. 9, the deflection angle and the azimuthal angle are determined by two random numbers ξ_θ and θ_φ:

$$\varphi = 2\pi \xi_\varphi \quad \text{and}$$

$$\cos\cos\theta = \{\frac{1}{2g}\left\{1 + g^2 + \left[\frac{1-g^2}{1-g^2+2g\xi_\theta}\right]^2\right\} \quad if\ g > 02\xi_\theta \quad if\ g = 0. \quad (23)$$

The Monte Carlo method becomes very accurate as the number of photon increases, and is often compared to other light-material interaction models to verify their accuracy.

Dipole BRDF and Multipole BSSRDF is introduced by Jensen et al. [15] to simulate exact subsurface light transport inside a highly scattering semi-infinite medium. It solves directly the radiative transfer theory by applying the diffusion approximation. The diffusion theory says that the radiance L is considered to be largely isotropic, by expressing L into a set of spherical harmonics, only the isotropic and 1st-order anisotropic terms need to be used, this gives

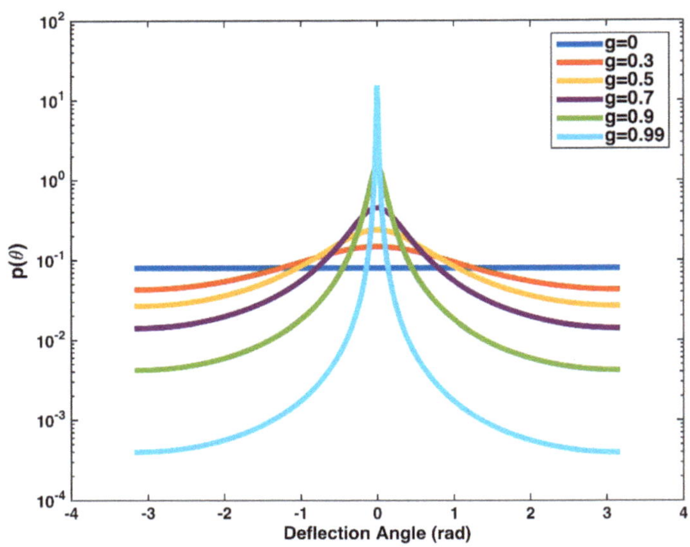

Fig. 9 Henyey-Greenstein phase function

$$L(x, \vec{\omega}) = \frac{1}{4\pi}\phi(x) + \frac{3}{4\pi}\vec{\omega} \cdot \vec{E}(x). \tag{24}$$

This is true even if the light source distribution and the phase function is highly anisotropic, because as the number of scattering event increases, the light distribution tends toward uniform. By substituting 24 into 8 and 7, we obtain the classical equation

$$D\nabla^2\phi(x) = \sigma_a\phi(x) - Q_0(x), \tag{25}$$

and a simple relation between the scalar and vector irradiance

$$\vec{E}(x) = -D\vec{\nabla}\phi(x). \tag{26}$$

where $Q_0(x) = \int_{4\pi} Q(x, \vec{\omega})d\Omega$ is the 0th-order source terms. $\sigma_t' = \sigma_a + \sigma_t(1 - g)$ is the reduced extinction coefficient and $D = 1/3\sigma_t'$ is the diffusion constant. The diffusion Eq. 25 has a simple solution in case of a single isotropic point source in an infinite medium, the well-known Green function

$$\phi(x) = \frac{\Phi}{4\pi D}\frac{e^{-\sigma_{tr}r(x)}}{r(x)}. \tag{27}$$

where Φ is the power of point light source; $r(x)$ is the distance between measuring point x and the point source; $\sigma_{tr} = \sqrt{3\sigma_a\sigma_t'}$ is the effective transport coefficient.

If the medium is a bounded domain, the inward diffuse flux at the medium surface is equal to the reflected outward flux

$$\int_{2\pi^-} L(x, \vec{\omega})(\vec{n}\&\vec{\omega})d\Omega = F_{dr}\int_{2\pi^+} L(x, \vec{\omega})(n_+.\vec{\omega})d\Omega. \tag{28}$$

where the sign $2\pi^-$ and $2\pi^+$ signify the hemi-sphere with normal pointing inward $\vec{n}_-$ and outward $\vec{n}_+$ the medium at point x. By applying the diffusion approximation (Eqs. 24 and 26), we reach the diffuse BSSRDF Rd

$$R_d(r) = -D\left(\frac{\vec{n} \cdot \vec{\nabla}\phi(x_o)}{d\Phi_i(x_i)}\right). \tag{29}$$

where $r = \|x_o - x_i\|$ is the distance between incoming and outgoing point of light. In case of a pencil beam light source, a dipole approximation is made. The light source is approximated by a pair of light sources: a real source located below the medium surface at distance z_r, the other one, virtual light source located above the surface at distance z_v, as shown in Fig. 10a. The reflectance is simply the sum of each point source distribution

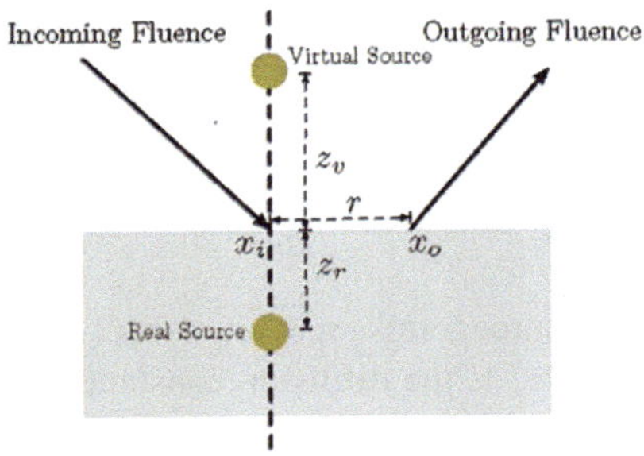

(a) Dipole approximation

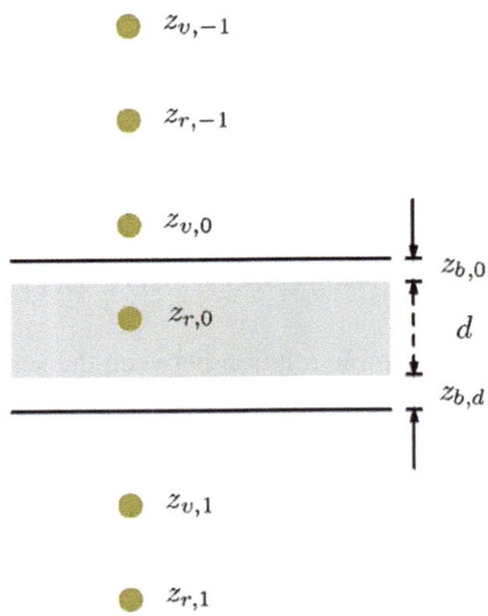

(b) Multipole approximation

Fig. 10 Dipole and multipole approximation

$$R_d(r) = \frac{\alpha'}{4\pi}\left[\frac{z_r(1+\sigma_{tr}d_r)e^{-\sigma_{tr}d_r}}{d_r^3} + \frac{z_v(1+\sigma_{tr}d_v)e^{-\sigma_{tr}d_v}}{d_v^3}\right]. \tag{30}$$

where $\alpha' = \sigma_s'/\sigma_t' = \sigma_s(1-g)/\sigma_t'$ is the reduced albedo. z_r and z_v are distances from the entering point to the light sources, which can be calculated $z_r = 1/\sigma_t'$ and $z_v = z_r + 4AD$ with $A = (1+F_{dr})/(1-F_{dr})$ a function of average Fresnel diffuse reflectance. d_r and d_v are the distances from the outgoing point x_o to the real and virtual light sources, respectively.

As the dipole BSSRDF works only for semi-infinite material, it fails in case of thin layer, because no transmission is taken into account. Jensen et al. extended this model to an arbitrary number of translucent layers with different thickness and optical properties, called multipole BSSRDF [24]. In case of a finite medium with thickness d, they substituted the contribution of 2 point sources by 2n + 1 pairs of point sources, as shown in Fig. 10b. This results a reflectance and transmittance in the following form:

$$\begin{aligned} R(r) &= \sum_{i=-n}^{n}\left[\frac{z_{r,i}\left(1+\sigma_{tr}d_{r,i}\right)e^{-\sigma_{tr}d_{r,i}}}{4\pi d_{r,i}^3} + \frac{z_{v,i}\left(1+\sigma_{tr}d_{v,i}\right)e^{-\sigma_{tr}d_{v,i}}}{4\pi d_{v,i}^3}\right], \\ T(r) &= \sum_{i=-n}^{n}\left[\frac{(d-z_{r,i})\left(1+\sigma_{tr}d_{r,i}\right)e^{-\sigma_{tr}d_{r,i}}}{4\pi d_{r,i}^3} + \frac{(d-z_{v,i})\left(1+\sigma_{tr}d_{v,i}\right)e^{-\sigma_{tr}d_{v,i}}}{4\pi d_{v,i}^3}\right]. \end{aligned} \tag{31}$$

where $z_{r,i} = 2i\left(d + z_{b,0} + z_{b,d}\right) + z_r$ and $z_{v,i} = 2i\left(d + z_{b,0} + z_{b,d}\right) - z_r - 2z_{b,0}$ with $z_{b,x} = 2A_x D$ the extrapolation distance is calculated separately for both top ($x = 0$) and bottom ($x = d$) surfaces.

3.3 *Skin Characterization Based on Light-Skin Interaction Model*

Skin is a complex multi-layered material consists of various components, see Fig. 11a.

In biophysical-based models (refer to Kubelka-Munk model and multipole BSSRDF here), the reflectance and transmittance are expressed by a function of biological parameters of skin: σ_a and σ_s, d. And a multi-layered skin model is made as shown in Fig. 11b. Skin is supposed to be a plane-parallel material with different optical parameters, the black circles in the diagram represent normally distributed absorbing and scattering constituents. For each skin layer, the optical parameters can be written as the sum of contributions of each constituent:

$$\sigma_a = \sum_i f_i \sigma_a^i, \sigma_s = \sum_i f_i \sigma_s^i. \tag{32}$$

where σ_i and σ_i represent the absorption and scattering coefficients of constituent i. Generally, the skin is considered as a three-layered material, known as epidermis, dermis and hypodermis, from the outmost to the deepest. We can select different combinations of pigments that are distributed in each layer:

- Epidermis contains two types of melanins (eumelanin and pheomelanin), β-carotene, bilirubin etc.;
- Dermis contains two types of hemoglobins (oxy- and deoxy-hemoglobin), beta-carotene, bilirubin etc.;

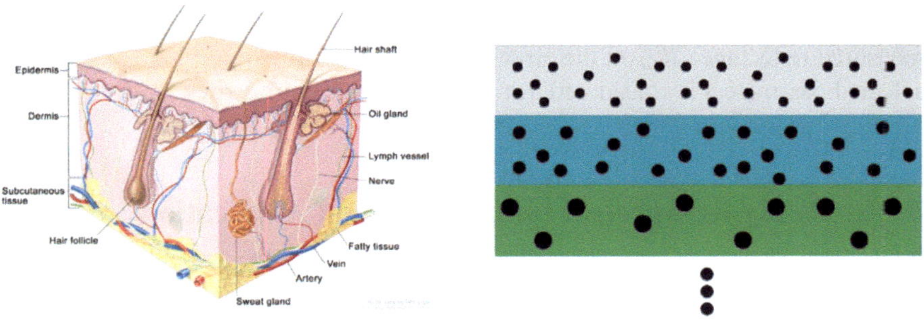

(a) Anatomy of human skin (b) Abstract human skin model

Fig. 11 Diagram illustrating anatomy of human skin and abstract skin model

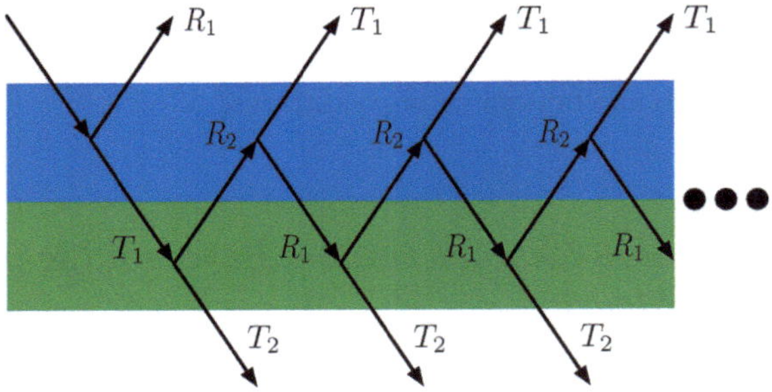

Fig. 12 Light interaction with different layers of skin

- Hypodermis is considered as an opaque layer, all light transmitted into this layer is absorbed.

Light in the skin will be reflected and transmitted between epidermis and dermis, see Fig. 12, in order to take this into account, the following expression is used to calculate the skin reflectance R and transmittance T for the upper 2 layers:

$$R = R_1 + \sum_{i=0}^{\infty} T_1(R_2R_1)^i R_2T_1 = R_1 + \frac{T_1R_2T_1}{1 - R_2R_1}, T = \sum_{i=0}^{\infty} T_1(R_2R_1)^i T_2 = \frac{T_1T_2}{1 - R_2R_1}. \tag{33}$$

where scalars R_1, R_2, T_1 and T_2 are overall reflectances and transmittances for layer 1 and 2, respectively. If the reflectance and transmittance of each layer is represented by 2D functions, the skin reflectance and transmittance are also 2D functions, and can be computed through the sum of 2D convolutions

$$\begin{aligned} R(x, y) &= R_1 + T_1 * R_2 * R_1 * R_2 * T_1 + T_1 * R_2 * R_1 * R_2 * R_1 * R_2 * T_1 + \ldots T(x, y) \\ &= T_1 * T_2 + T_1 * R_2 * R_1 * T_2 + T_1 * R_2 * R_1 * R_2 * R_1 * T_2 + \ldots \end{aligned} \tag{34}$$

where $(f * g)(x, y) = f(x, y) * g(x, y) = \int_{-\infty}^{\infty} f(x', y')g(x - x', y - y')dx'dy'$.

One the biological parameters and the skin model is selected, Fig. 13 illustrates a framework of the skin biological parameters extraction. The skin parameters are selected by comparing the error between modelled reflectance R_{mo} between the measured reflectance R_{me}, if the error is within the proposed threshold ε, this set of selected parameters is considered as the skin biological properties.

4 Multi-Hyper-Spectral Imaging Related Applications

In medical diagnosis, determining skin optical parameters has important meaning. Some diseases, for example, the jaundice is a yellowish pigmentation of the skin, which is caused by high bilirubin concentrations in blood [26]. Jaundice is often seen in liver disease such as hepatitis or liver cancer. It may also indicate leptospirosis or obstruction of the biliary tract. But yellow discoloration of the skin, especially on

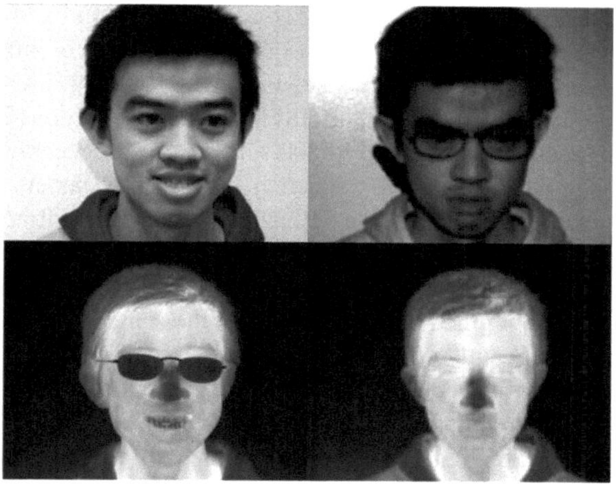

Fig. 13 Face images from Laval University multispectral face database (visible, NIR—Near Infra-Red, MWIR—Med Wave Infra-Red and LWIR—Long Wave Infra-RED) [25]

palms and the soles, is due to carotenemia—a harmless condition [27]. This needs to be differentiate from jaundice. Another example is the determining the differences between skin cancer and lesions. In clinical diagnosis, there is still challenges in precisely differentiate lesions and malignant melanoma. Most dermatologists still rely on their practical experience in visual evaluations. Pavlova et al. [28] studied the differences among healthy skin, skin cancer and lesions spectral reflectance. Their study shows that there exist distinguishable differences between different state of the skin. Computer-assisted diagnosis with skin biological parameters will significantly increase the diagnostic accuracy.

Cosmetics Cosmetic industry is highly related to skin biology, as a great number of products are developed to affect the skin appearance and make the skin looks better. These cosmetics will change the concentration of certain skin constituents. For example, the whitening cream, it contains chemical substances in an attempt to lighten skin tone or provide an even skin complexion by lessening the concentration of melanin. Specific zones of abnormally high pigmentation of skin, such as moles and birthmarks may be depigmented to match the surrounding skin, after applying these kind of products. The biophysical-based light-skin interaction models provide a quantitative evaluation and study of cosmetics by comparing the melanin concentration before and after a period of applications. In cosmetic development, combinations of biological process simulations and light-skin interactions can provide a direct visual result for the case of application of certain chemical substances.

Still in search of more robust and effective biometrics approaches, the use of MHSIs has been advocated. For instance, many studies proposed to use the infrared spectra. Here, known techniques dedicated to visible 2D face recognition, such as Eigenfaces or Fisherfaces, are also used with infrared images. In [22], based on thermal face images, facial physiological information is extracted from high temperature regions and used in IR face recognition. In [25], authors propose a comparative study of Multispectral Infrared Face Recognition. They also introduce the use of non-linear dimensionality reduction techniques and a probabilistic Bayesian based face recognition technique. Experiments conducted on Equinox Multimodal and Laval University Multispectral Face Databases show a reduction of intrapersonal variation, leading to interesting performance, see Fig. 13.

Pan et al. proposed the HSI for facial recognition [29]. They applied near-infrared hyperspectral imaging to 200 subjects for the recognition of their faces. The hyperspectral images were collected using a CCD camera equipped with a liquid crystal tunable filter to provide 31 bands over the near-infrared (0.7–1.0 μm). They argued that hyperspectral imaging provides useful discriminants that cannot be obtained by other imaging methods. They demonstrated experimentally that spectral measurements over the near-infrared allow the sensing of subsurface tissue structure which is significantly different from person to person, but relatively stable over time. Also, they established that the local spectral properties of human tissue are nearly invariant to face orientation and expression which allows hyperspectral discriminants to be used for recognition over a large range of poses and expressions.

Soft biometrics like gender, ethnicity and age play an important role in various domains. Human can easily distinguish people of different gender, ethnicity, and age according to skin appearance; it is still a challenge for computer vision. Gender, ethnicity and age classifications based on facial images have been widely studied, features such as LBPs [30], SIFT [31] etc. show no explicit relations with skin biological characteristics, and yield satisfying results only in gender classification. Techniques based on light-skin interaction models are rarely investigated. Malskies et al. [32] applied a 4-layered skin model consisting of epidermis, upper papillary dermis, lower papillary dermis and reticular dermis. The epidermis contains melanin and is supposed to be non-scattering; hemoglobin rations of the 2 papillary dermis layers are related to the proportion of their thickness. By applying the Kubelka-Munk model, the overall skin reflectance is expressed by a function of 5 biological parameters. The parameters are used directly as the input feature of ethnicity classification, and reached an overall classification rate of 96.52%. Wei et al. proposed a classification scheme [33] based on BRDF albedo. They selected 5 facial regions (philtrum, forehead, nose, cheek and chin); then gender, skin type and age classifications using albedo distribution histograms is performed; finally, a weighted sum fusion technique is applied to improve the classification rate. The classification results using a biophysically based skin model reached 96.88, 95 and 90.63% for gender, skin type and age respectively.

Although biometrics authentication and identification stand out with their reliability and convenience, there are still drawbacks that allows circumvention of surveillance with spoofing attacks. Spoofing attack is defined as a counterfeit biometric that is used to circumvent a biometric sensor [34]. For instance, in order to achieve authentication, one can display high-resolution photos on displaying medias in front of a biometric sensor. Most spoofing attacks are based on displayed photos/videos or printing artifacts, the success of the attacks is highly relying on their qualities and the type of recognition system. While there are a multitude of anti-spoofing techniques, the most effective of them are based on MHSI. A first category of approach, is based on liveness verification. For instance, eye blink detection, depth verification. Simple to implement, these approaches fail in case of spoofing attacks from videos, and high quality masks with eyes and mouth regions carved out. In fact, a 3D printed wearable mask manufactured by That'sMyFace.com is a great success in spoofing attacks, and claims to have a 100% success rate in attacking image-based face recognition systems. In order to overcome this type of attacks, Kim et al. [35] and Zhang et al. [36] both distinguished skin by applying directly skin reflectance obtained with Lambertian model, where the materials are assumed to be opaque. This passes the problem from colorimetry to radiometry, as different spectral energy distributions can produce the same "color" to human perception as well as the RGB or grayscale images. Although these methods can yield satisfying results in identifying materials different from skin, the main drawback is that these methods take reflectance from only a few distinguishable wavelengths mostly in NIR range, and are particularly designed to classify skin from silicon and papers.

To distinguish skin from other materials, we examine in the next section three different approaches [37, 38]. Two of them use directly the reflectance spectra while the third one performs an indirect application.

4.1 Direct Application of Reflectance Spectra

The reflectance at 850 nm is highly discriminative and allows to separate the skin from silicon or latex, while the visible spectrum range, 380–720 nm, is used for texture based individual identification or authentication. In this case the whole is considered as a direct application of multi spectral imaging. An alternative solution is to use the reflectance in all visible spectrum range as the input for anti-spoofing and identification or authentication. In the following, we introduce two direct applications and one indirect application of the reflectance spectra for skin / non-skin classification. The reflectance at 516, 540, 564 and 576 nm which produce discriminative signatures, known as the "W" pattern can be recognized in a direct way. This "W" pattern is due to absorption of oxy-hemoglobin, Fig. 14 illustrates some absorption coefficients of pigments in human skin.

Also, as a second descriptor, V_f, only a few number of selected spectra describe a 3 "true-false" patterns:

$$\begin{aligned} V_f = (v_1\, v_2\, v_3) = (&\text{is } R(516\text{ nm}) > R(540\text{ nm})\text{is } R(540\text{ nm}) \\ &> R(564\text{ nm})\text{is } R(564\text{ nm}) > R(576\text{ nm})) \end{aligned} \quad (35)$$

4.2 Indirect Application of Reflectance Spectra

The indirect application of the skin reflectance spectra consists of estimating intra structural parameters of the skin tissues, for instance a simple 2-layered skin with different pigments, for skin non skin classification: the epidermis contains 2 types of

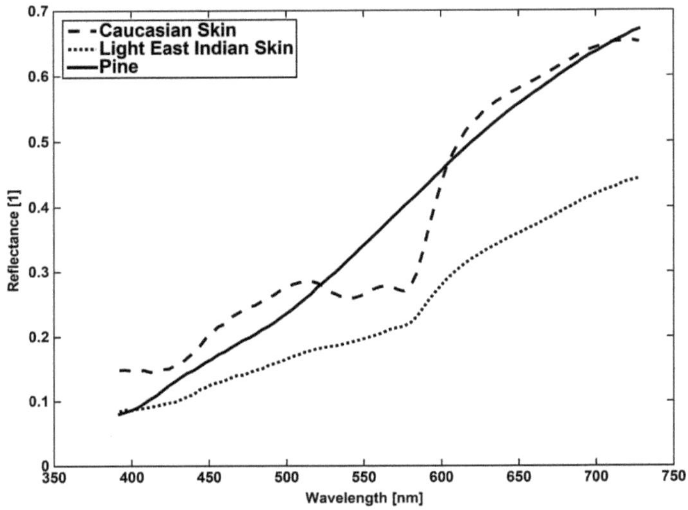

Fig. 14 Comparison of pine and different types of skin reflectance

melanins (eumelanin and pheomelanin), and has a constant thickness of 0.25 mm; the dermis contains 2 types of hemoglobin (oxy- and deoxyhemoglobin), and is considered to be infinitely thick. The absorption coefficients σ_a^{epi}, σ_a^{derm} of epidermis and dermis are given by:

$$\sigma_a^{epi} = f_{me}\left(b_{me}\sigma_a^{me} + (1 - b_{me})\sigma_a^{pm}\right) + (1 - fme)\sigma_a^{dp} \tag{36}$$

$$\sigma_a^{derm} = f_{hg}\left(b_{oh}\sigma_a^{oh} + (1 - b_{oh})\sigma_a^{dh}\right) + (1-fhg)\sigma_a^{dp}$$

where f_{me} is the volume fraction of 2 types of melanin in the epidermal layer, b_{me} controls the ratio of eumelanin to pheomelanin, f_{hg} is the volume fraction of hemoglobin in the dermal layer, r_{oh} represents the ratio of oxy- to deoxyhemoglobin and is fixed to 0.75. The σ_a^{em}, σ_a^{pm}, σ_a^{oh}, σ_a^{dh} and σ_a^{dp} refer to absorption coefficients of eumelanin, pheomelanin, oxy-hemoglobin, deoxy-hemoglobin and depigmented skin tissue respectively, as illustrated in Fig. 15.

The epidermal scattering coefficients σ_s^{epi} can be calculated by the sum of Rayleigh and Mie scattering coefficients, the dermis is thought to have 1.5 times higher scattering coefficients than epidermis. The overall reflectance profile can be expressed as follows by accounting for inter-layer reflections:

$$R = R_1^+ + T_1^+ * R_2^+ * T_1^- + T_1^+ * R_2^+ * R_1^- * R_2^+ * T_1^- \dots \tag{37}$$

where R_i and T_i represent reflectance and transmittance profiles of *i-th* layer, the superscript + and − indicate that the profile is in the case of downward or upward propagation respectively, and can be computed simply by swapping the refractive

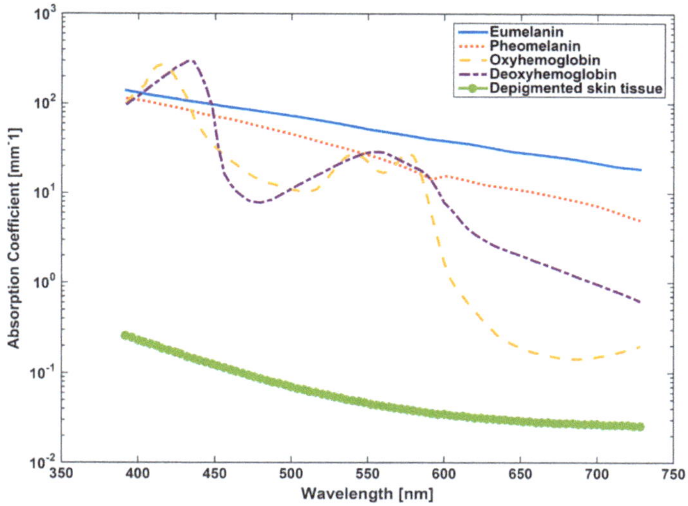

Fig. 15 Spectral absorption coefficients of hemoglobin, melanins and depigmented skin tissue

index of the upper and lower materials. The parameters are extracted by finding the set of parameters that minimizes

$$\varepsilon = \sum_{\lambda} |R_{measured}(\lambda) - R_{modeled}(\lambda)| \tag{38}$$

At last the parameter set (f_{me}, b_{me}, f_{hg}) forms the descriptor. To evaluate the proposed approaches, the reflectance spectra measured in [39], have been used. This database contains 170 reflectance spectra: 30 of which are human skin reflectance spectra, the others are from various natural and man-made objects such as rocks, plants, and hair, and fabrics. In the first experiment, the skin classification with RBF-kernel SVM has been implemented. The classification is evaluated by a 5-fold cross validation, each sample is classified only once and will not present in both training and testing samples at the same time. Figure 4 show some misclassified reflectance spectra of skin (dot) and non-skin materials (line). The misclassified skin reflectance spectra samples have very similar shapes, they show no obvious "W" pattern; and for the misclassified non-skin materials, their spectra have significant increase in range 550–620 nm. Also the performance of the "W" pattern descriptor has been evaluated. A sample is classified as a skin, if and only if all values of the descriptor are true. The accuracy in classifying non-skin material reaches 100%, this could have important meaning in anti-spoofing attacks. As this descriptor offers a simple yet accurate method in classifying non-skin materials with only a little loss of accuracy in detecting human skin comparing to using the whole visible spectra, the reflectance spectra measures of 4 discrete wavelengths are needed. The final evaluation uses the synthesized optical parameters by fitting the measured reflectance spectra to the

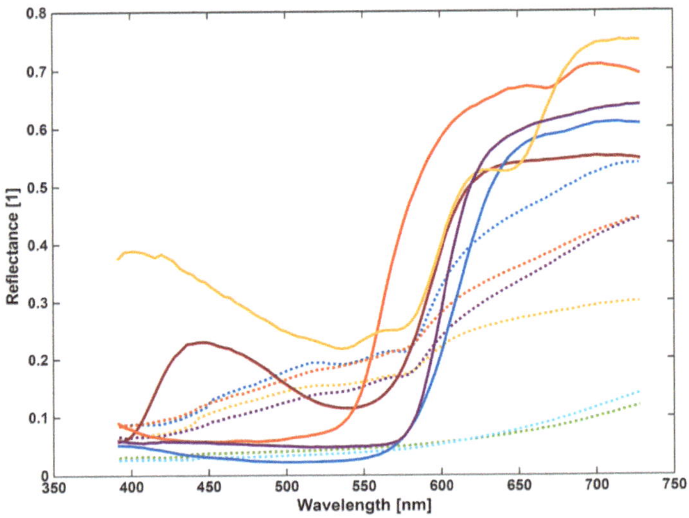

Fig. 16 Misclassified skin samples

proposed 2-layered skin model. In the fitting step, all parameters f_{me}, b_{me} and f_{hg} are assumed to vary in interval [0, 1]. The classifier and experiment setup remain the same with the first experiment. As expected, the overall classification result reaches an accuracy of 96.4%, and Fig. 5 shows the output of descriptors. Here the data is orthogonally projected onto the two dimensional $f_{me} - f_{hg}$ plane, that are most discriminative parameters. We can notice that most of the skin samples (circles) are centered at the bottom-left region, while the non-skin samples (plus signs) are distributed mostly at the top and bottom line. The reason may quite simple, because the light-material interaction model, where the reflectance spectra are fitted to, is intended to explain the light-skin interaction only, which cannot precisely describe interactions with non-skin materials. Human skin consists of various compositions can be found only in vivo such as melanin and hemoglobin, non-skin materials consequently cannot fit well to the proposed skin model, even if it is a very simple one (Figs. 16 and 17).

5 Conclusion

In this chapter, we examined hyperspectral imaging and analyzing approaches dedicated to biometrics. Also, we introduced basic concepts of optics and radiative transfer theory, then presented several light-skin interaction models allowing indirect characterization of the skin. Skin estimated features can be used for applications in biometrics such as identification, authentication, soft-biometrics (sex, age and ethnic

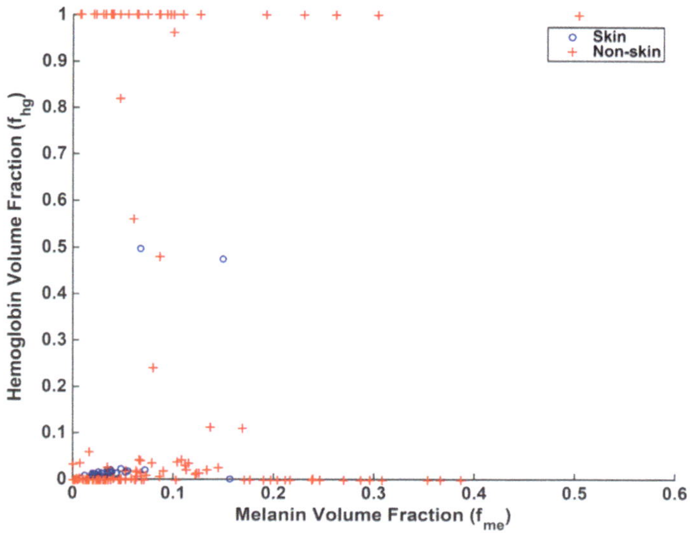

Fig. 17 Data distribution on orthogonally projected fme–fhg plane

classification), and skin/non skin classification as countermeasure to anti-spoofing attacks.

None of the models is perfect, each model has its own advantages and disadvantages. The specular reflection model and Lambertian model consider only one form of reflection, but this results its simplicity and low cost in computation, which can be widely used in computer and video games. The Blinn-Phong model can describe different types of surfaces ranging from smooth to rough, and considers ambient light into account. The Torrance-Sparrow model can yield plausible results under different conditions, this model can be combined with dipole or multipole BSSRDF by replacing the surface Fresnel reflectance by a Torrance-Sparrow reflectance. This allows these two models to express skin oiliness by a roughness parameter. But the parameters of both Blinn-Phong and Torrance- Sparrow models cannot be measured directly, users need to experiment until they get a satisfying result; in some angles, these two models do not obey the energy conservation law, this results unrealistic outputs. Although the Kubelka-Munk model describes the subsurface light transport, and can be extended to multi-layered materials; it is restricted to one-dimensional tasks, no lateral scattering is occurred. The Monte Carlo method is very accurate, and is often used to evaluate other models. But it needs a very large amount of simulations to acquire accurate results, which is very time and computational consuming; and is very difficult in modelling complex scenes or inhomogeneous materials. The dipole BSSRDF describes accurate subsurface light transport, but it fails when the material is not semi-infinite. The multipole BSSRDF can obtain very accurate results comparing with Monte Carlo simulation, and is much less time consuming and computationally expensive; it fails when being applied to very thin materials.

Finally, at the last part of this chapter, we detailed some applications or perspectives of applications of these light-skin interaction models. Except in computer graphics, applications of these models are rarely investigated. Comparing with other non-invasive techniques such as OCT, MRI etc., approaches based on light-skin interaction models have great potential. Because the acquisition devices are relative at low cost, and some devices introduced in the previous section can yield high spatial or spectral resolution to extract accurate biological parameters.

References

1. Lu, G., Fei, B.: Medical hyperspectral imaging: a review. J. Biomed. Opt. **19**(1), 010901 (2014)
2. Hagen, N., Kudenov, M.W.: Review of snapshot spectral imaging technologies. Opt. Eng. **52**(9), 090901 (2013)
3. Gat, N.: Imaging spectroscopy using tunable filters: a review. Proc. SPIE **4056**, 50–64 (2000)
4. Lyot, B.: Optical apparatus with wide field using interference of polarized light. C. R. Acad. Sci. (Paris) **197**, 1593 (1933)
5. Šolc, I.: Birefringent chain filters. J. Opt. Soc. Am. **55**(6), 621–625 (1965)
6. Shriyan, S.K.: Tunable electro-optic thin film stack for hyperspectral imaging. Ph.D. Thesis, Drexel University (2011)
7. Shriyan, S.K., Schundler, E., Schwarze, C., Fontecchio, A.K.: Electro-optic polymer liquid crystal thin films for hyperspectral imaging. J. Appl. Remote Sens. **6**(1), 063549 (2012)
8. Golub, M., Menachem, N., Amir, A., Kagan, A., Zheludev, V., Malinsky, R.: Snapshot spectral imaging based on digital cameras. U.S. patent US 20130194481 A1 (2013)
9. Gehm, M.E., John, R., Brady, D.J., Willett, R.M., Schulz, T.J.: Single-shot compressive spectral imaging with a dual-disperser architecture. Opt. Express **15**(21), 14013–14027 (2007)
10. Golub, M.A., Nathan, M., Averbuch, A., Lavi, E., Zheludev, V.A., Schclar, A.: Spectral multiplexing method for digital snapshot spectral imaging. Appl. Opt. **48**(8), 1520–1526 (2009)
11. Habel, R., Kudenov, M., Wimmer, M.: Practical spectral photography. EUROGRAPHICS **31**(2) pt2, 449–458 (2012)
12. Hegyi, A., Martini, J.: Hyperspectral imaging with a liquid crystal polarization interferometer. Opt. Express **3**(22), 28742–28754 (2015)
13. Baranoski, G.V.G., Krishnaswamy, A.: Light & Skin Interactions, Simulations for Computer Graphics Applications. Morgan Kaufmann (2010)
14. Chandrasekhar, S.: Radiative Transfer. Courier Corporation (2013)
15. Jensen, H.W., Marschner, S.R., Levoy, M., Hanrahan, P.: A practical model for subsurface light transport. SIGGRAPH, 511–518 (2001)
16. Lambert, J.H., Anding, E.: Ostwalds Klassiker der exakten Wissenschaften. Lamberts Photometrie (Photometria, sive De mensura et gradibus luminis, colorum et umbrae) (1760). W. Engelmann, 1892
17. Phong, B.T.: Illumination for computer generated pictures. Commun. ACM **18**, 311–317 (1975)
18. Torrance, K.E., Sparrow, E.M.: Off-specular peaks in the directional distribution of reflected thermal radiation. J. Heat Trans. **88**, 223–230 (1966)
19. Sparrow, E.M., Torrance, K.E.: Theory for off-specular reflection from roughened surfaces. JOSA **57**, 1105–1112 (1967)
20. Kubelka, P., Munk, F.: Ein beitrag ztlr optik der farbanstriche. Z. Teeh. Phys. **12**, 593–601 (1931)
21. Wang, L., Jacques, S.L., Zheng, L.: MCML—Monte Carlo modeling of light transport in multi-layered tissues. Comput. Methods Progr. Biomed. **47**, 131–146 (1995)
22. Wang, L., Jacques, S.L.: Optimized radial and angular positions in Monte Carlo modeling. Med. Phys. **21**, 1081–1083 (1994)

23. Shirley, P.: Nonuniform random point sets via warping. In: Graphics Gems III, pp. 80–83. Academic Press Professional Inc. (1992)
24. Donner, C., Jensen, H.W.: Light diffusion in multi-layered translucent materials. ACM Trans. Graph. **24**, 1032–1039 (2005)
25. Akhloufi, M., Bendada, A.: Multispectral infrared face recognition: a comparative study. In: 10th International Conference on Quantitative InfraRed Thermography, pp. 27–30. Québec (Canada) (2010)
26. Click, R., Dahl-Smith, J., Fowler, L., DuBose, J., Deneau-Saxton, M., Herbert, J.: An osteopathic approach to reduction of readmissions for neonatal jaundice. Osteopath. Family Phys. **5**, 17–23 (2013)
27. Arya, V., Grzybowski, J., Schwartz, R.A.: Carotenemia. Cutis **71**(6), 441–2, 448 (2003)
28. Pavlova, P., Borisova, E., Petkova, E., Avramov, Troyanova, P.: Investigation of Relations Between Skin Cancer Lesions' Images and Their Reflectance and Fluorescent Spectra (2011)
29. Pan, Z., et al.: Face recognition in hyperspectral images. IEEE Trans. Pattern Anal. Mach. Intell. **25**(12), 1552–1560 (2003)
30. Huang, D., Shan, C., Ardabilian, M., Wang, Y. Chen, L.: Local binary patterns and its application to facial image analysis: a survey. IEEE Trans. Syst. Man. Cybern. Part C (Appl. Rev.) **41**, 765–781
31. Lowe, D.G.: Distinctive image features from scale-invariant keypoints. Int. J. Comput. Vision **60**, 91–110 (2004)
32. Malskies, C.R., Eibenberger, E., Angelopoulou, E.: The recognition of ethnic groups based on histological skin properties. In: Proceedings of Vision, Modeling, and Visualization, pp. 353–360 (2011)
33. Chen, W., Ardabilian, M., Zahouani, H., Zine, A.: Gender, skin type and age classifications using skin reflectance-based descriptor. Rap. Tech. (2014)
34. Nixon, K.A., Aimale, V., Rowe, R.K.: Spoof detection schemes. In: Handbook of Biometrics. Springer US, Boston, MA, pp. 403–423 (2008)
35. Kim, Y., Na, J., Yoon, S., Yi, J.: Masked fake face detection using radiance measurements. JOSA A (2009)
36. Zhang, Z., Yi, D., Lei, Z., Li, S.Z.: Face liveness detection by learning multispectral reflectance distributions. In: Face and Gesture, pp. 436–441(2011)
37. Chen, W., Ardabilian, M., Zin, A., Zahouani, H.: Reflectance spectra based skin and non-skin classification. Proc. ICIP **2015**, 755–759 (2015)
38. Ardabilian, M., Zine, M., Chen, W.: Procédé pour distinguer automatiquement une peau d'un être humain d'un leurre inanimé de peau humaine, Patent, WO2017051116A1
39. Vrhel, M.J., Gershon, R., Iwan, L.S.: Measurement and analysis of object reflectance spectra. Color Res. Appl. **19**(1), 4–9 (1994)
40. Hecht, E.: Optics. Addison-Wesley Longman (2002)

Imaging for Hidden Biometrics

Delphine Maugards and Amine Nait-ali

Abstract In this chapter visible biometrics is highlighted though a classical review on face recognition, including some known approaches from the literature. Therefore, it will be evoked some multispectral approaches such as Near Infrared (NIR) and Infrared (IR), and in the same context, deep hidden biometrics using X-ray imaging will be considered. Finally, a promising and safe MRI biometrics is described through some recent advances in brain biometrics, considered as a robust and safe modality. The purpose of using some hidden biometrics imaging techniques consists basically into preventing potential attacks.

1 Introduction

Biometrics in the visible light has been highly investigated in the literature. In fact, this conventional biometrics does not require sophisticated equipment to represent the region of interest since, generally the information is captured through a simple RGB camera. As an example, face recognition, gait analysis, hand shape recognition or ear recognition may be classified in this category. The problem with this kind of modalities is their sensitivity to acquisition conditions, in particular, to illumination. Very often, illumination cannot be controlled. Therefore, this may directly impact the performance of the biometric system. When it comes to biometrics in the visible light, another problem can raise, namely, the sensitivity of this modality to attacks. If we consider for example face recognition, spoofing can be easily achieved through some simple techniques (e.g. photography, video, etc.). Of course, specific recognition algorithms can be designed to detect some attacks attempt, but the performances of such systems may be drastically limited.

Within this context, hidden biometrics can be employed to overcome this type of issues. For instance, instead of using the visible light, one can use the multispectral imaging such as Near Infrared (NIR) or Infrared (IR) to visualize some hidden

D. Maugards · A. Nait-ali (✉)
Université Paris-Est, LISSI, UPEC, 94400 Vitry sur Seine, France
e-mail: naitali@u-pec.fr

A. Nait-ali (ed.), *Hidden Biometrics*, Series in BioEngineering,
https://doi.org/10.1007/978-981-13-0956-4_9

layers of the skin. This strategy can even be pushed further through other imaging techniques, including X-ray and MRI.

This chapter is organized as follows: in the next section, visible biometrics is highlighted though a classical review on face recognition, including some known approaches from the literature and commercial softwares. Therefore, we evoke some multispectral approaches such as Near Infrared (NIR) and Infrared (IR). By keeping the same logic, deep hidden biometrics using X-ray imaging is presented. Finally, a promising and safe MRI biometrics is presented by describing some recent advances in brain biometrics, considered as a robust and a safe modality.

2 Biometrics in the Visible Light: A Review on Face Recognition

In this section, some common face recognition approaches will be reviewed. Basically, these approaches have been used for face recognition acquired in the visible light, while such algorithms can be adapted for other acquisition systems, including those requiring specific types of imaging.

2.1 PCA/EigenFaces Based Approach

One of the classical approaches of face recognition techniques consists in extracting nodes or landmarks from an image using PCA combined to Eigenfaces, in order to reduce the dimension of the features to orthogonal and uncorrelated components. The result is a dramatic downsizing of the raw data size into a 1D weighted feature vector, which makes the algorithm convenient to process when the face has to be compared to a large set of images. Methods may combine in a first step the detection of local positions of the facial keypoints on the image, then a cascade of convolutional neural networks to compute the extracted features.

The main limitation of this technique is that it requires a full frontal face to be acquired and compared to similar frontal images. Affine transformation and other automatic shape alignment step may compensate this limitation.

2.2 Discriminant Analysis

The popular Support vector Machine (SVM) is a binary or multi-class algorithm in the category of the maximum margin classifier. It has been formulated to distinguish the disparities between faces of different people, or disparities between images of the same person. A single feature vector represents the whole face image before getting processed for recognition and identification. The classification can be made robust to face rotation changes up to 40% when combined with local facial component classification.

Other traditional algorithms, such as Linear Discriminant Analysis (LDA or Fisherface) and Quadratic Discriminant Analysis (QDA), are unsupervised classifications for linear problems that maximize the distance between the clusters that are found, while minimizing the distance within each cluster. For face recognition the projected image is compared to each projected projection in the training data. The face identification can be performed using KNN.

PCA can also be combined with LDA in a two-step analysis method, first with PCA for the projection of the face image, second with LDA to obtain a linear classifier when only a few samples per class are available.

Similarly to PCA and Eigenfaces in terms of dimensionality reduction, the results can be subject to overfitting, are limited to frontal faces and require approximately normal densities of clusters.

A full review and comparison of SVM and discriminant analysis with five linear methods has concluded that SVM was to be chosen over other methods.

2.3 EBGM Based Approaches

Elastic Bunch Graph Matching, or EBGM, is based on Gabor wavelet transform with nodes describing pixels on an elastic grid. A convolutional Gabor filter detects shapes while taking into account differences in illumination of the image and changes in face pose and expression. Its limitation lays in its dependency to accurate localization of landmarks on the face, requiring a combination with PCA/LDA.

2.4 Convolutional Neural Networks Based Approaches

Traditional computer vision algorithms have been increasingly supplanted by deep learning techniques and artificial neural networks. The main benefit of deep learning based algorithms is that they completely bypass the feature extraction and dimensionality reduction steps. We will focus on Convolutional Neural Networks (CNN) as a subset of feedforward neural networks used for computer vision. A CNN learns hierarchical features through multiple layers of convolutions and max-pooling layers, acting as a complex feature extractor. Convolution means not all the nodes in all the layers are connected to each other, in order to reduce the number of parameters (or weights) to calculate and optimize for each filter.

The higher level layers, or the final layers, are fully-connected in order to act as the final non-linear classifier. Transfer learning consists in using pre-trained CNN and in re-training only the highest layers, to detect specific objects or faces not present in the pre-trained model for example. CNN have been used in standalone mode (VGG-

Table 1 Face recognition on LWF, Results, Taigman et al. [1]

	Approach	Error %
1	Principal Component Analysis (Eigenfaces)	60.2
2	Local Binary Pattern (LBP)	72.4
3	Deep CNN (AlexNet) + Eudid (Ladzy DL)	71
4	DeepFace by Facebook (Standard DL)	97.25

16, VGG-32, GoogLeNet), in combination with Gabor filters or in combination with Local Binary Patterns Histograms (LBPH, or LP+HOG).

Face recognition has been such a focus for both the industry and research labs that tremendous progress has been made over the last 5 years. Today more than 60 2D- and 3D-databases can be found online for face recognition testing. As an example, the following results were found on the LFW database when testing Facebook's DeepFace algorithm [1] (Table 1).

2.5 *Commercial Face Recognition*

Commercial face recognition applications have developed tremendously and it may be puzzling to know what algorithm was used behind the mainstream applications. The very latest results of tests conducted in March 2018 and published in May by the US Homeland Security Science and Technology Directorate, known as the Biometric Technology Rally held in Maryland (MdTF), provides the latest list of the top facial algorithms. Face and combined face and iris recognition systems were put in competition in a setup reproducing a standard security checkpoint. Criteria included a processing time of less than 5 s and a 95% user satisfaction rate.

Among the competing face recognition algorithms, we can mention:

- DeepFace or FaceDB, developed by Facebook: uses the triplet loss method, with anchor image (stored in a reference database), one positive (same person with a different angle), one negative (a different person). The training set needs only 10 positive pictures of each person, even hard to recognize (hat, glasses, etc.).
- Academia: developed in 2014 by researchers at Hong Kong University, it achieved facial identification scores of 98.52%.
- FaceNet/OpenFace, developed by Google, using a HOG face detector and artificial neural network reaching up to 100% accuracy.
- Face++, developed by Microsoft, IBM and China-based Megvii, made the headlines in 2018 after a MIT study pointed out its biased facial recognition results.
- Rekognition, developed by Arts Technica and Amazon, is able to recognize a hundred faces in a single group picture, however may suffer from an insufficient accuracy.
- LFIS, developed by Gemalto.

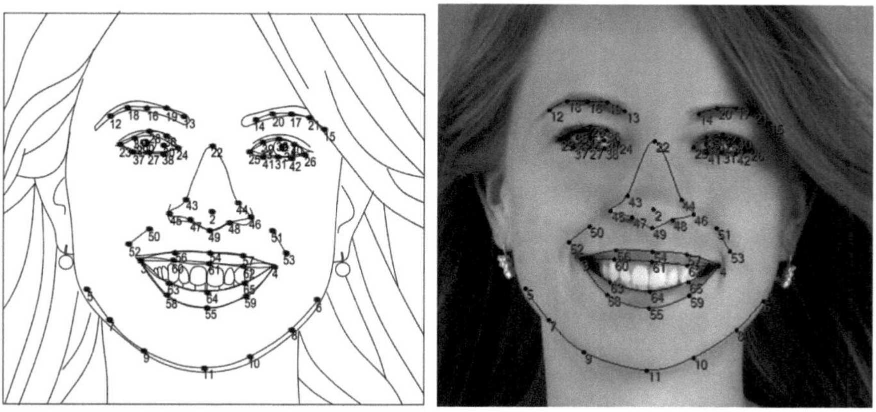

Fig. 1 Facial landmarks. Courtesy of Luxand, 2019

An additional focus of the industry has been on adapting the systems to mobile devices. Efforts are currently being made to adapt the algorithms to the limited processing power of a phone. In the future we may imagine such hidden biometric analysis performed through drones or mobile portals. Among the companies working in this field we may cite:

- BioID Facial Recognition
- True Key (created by Intel using Eigenfaces)
- FaceLock
- Luxand Face Recognition

As an example, Luxand FaceSDK (Fig. 1) is a face recognition library working for mobile, desktop and servers, that can detect facial features according to the images below. It employs a convolutional neural network of 2.7 million parameters, trained on a few million of faces, and resulting in a 256-dimensional vector for a specific facial image. The vector matching uses a cosine similarity.

3 Imaging Techniques for Hidden Biometrics

Although, face recognition is an acceptable common biometric modality, its performance can drastically decrease in low acquisition conditions such as: pose, illumination, ageing, facial expressions, etc. Moreover, face recognition can be vulnerable regarding attacks. For this reason, regardless of the ethical or the economic aspects, spoofing issue can be overcome though hidden biometrics which may use specific imaging techniques to characterise the whole face or a part of it or even some hidden anatomy including brain tissues. This may be achieved using spectroscopy (ultrasound, infrared, fluoroscopy, etc.), Computerized Tomography (CT), X-ray, MRI, etc.

3.1 Spectroscopic Techniques

Spectroscopy is the experimental subject concerned with the absorption, emission or scattering of electromagnetic radiation by atoms or molecules. The result is a spectrum, or a plot of measured electromagnetic radiation intensity versus wavelength. Within this context, one can highlight:

Optical spectroscopy: it includes luminescence (fluorescence, phosphorescence), infrared, Raman, UV.

Ultrasound: It is mostly known for their use during pregnancy to give parents a peek inside the womb. Other utilization of this technology have developed to evaluate and diagnose a range of medical conditions in cardiac, Ob/gyn, ophthalmic, vascular, abdominal, breast specialities and more. Ultrasound relies on sound waves to produce internal images of internal organs, soft tissue and blood vessels in a non-invasive way. In addition, the ultrasound technology relies on the skills of the sonography operator in order to correctly position the patient and record accurate images. This modality hasn't widely been used in security biometrics since the acquisition requires using gels between the sensors and the skin in order to ensure an optimal signal conduction.

Infrared (*IR*): it has been used in vein matching, or vascular pattern recognition (VPR) using thermal cameras. Primarily used in combination of the more established technique of fingerprinting or hand palm authentication, the vascular information revealed in our blood vessels provides a large volume of differentiating features for personal identification, enabling a high level of accuracy.

Near-infrared (*NIR*): light absorption by the red blood cells of the vein system is recorded as dark lines by a CCD camera behind a glass. A physical contact with the scanner is not needed. Vascular scanners are not affected by external light conditions or modifications outside the body.

If we consider face recognition, Near-infrared and infrared imaging have been used in numerous research work. For instance, the reader can refer to a review of methodologies and databases, provided in [5]. Figures 2 and 3 show respectively some samples from Terravic Facial IR and Databases and CBSR NIR Database.

3.2 X-Ray Imaging Approach

X-ray imaging is a candidate modality that has intensively used in the medical field and forensics, but rarely considered in biometrics. The main reason is the risk of radiation generated by associated systems and the impact that can have on health is still not under-control. In addition, this modality may impact users acceptability as well as the ethical aspect. Despite these limitations, recent applications deal with X-ray imaging when a high level of security is required. For instance, in some airports, full body scanners have been deployed for detection purpose. This concerns the following categories, namely:

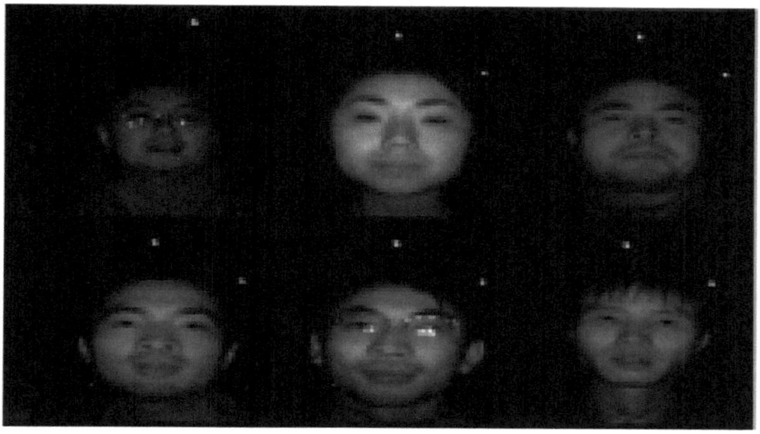

Fig. 2 Samples of Terravic Facial IR database

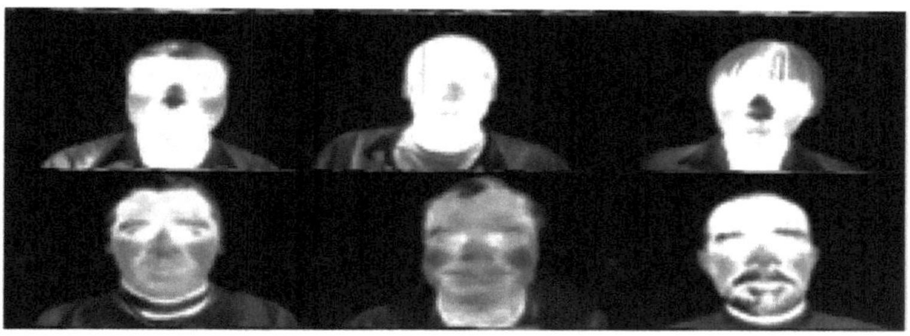

Fig. 3 Samples of CBSR NIR database

- Millimeter wave scanners: this scanner is the less invasive in its category. It uses non-ionizing radiations (Fig. 4).
- Backscatter X-ray: this scanner uses low dose radiation since it uses less than 0.0001 mSv/scan compared to a natural background radiation of around 3.1 mSv/year. It allows accurate detection of hidden objects (Fig. 4). In order to compare this amount of radiation with
- Through body X-ray: this category of scanners uses more radiation than the previous ones. It provides detailed body images.

Technically speaking, the scanners cited above may be adapted to biometric applications for person identification, but since the application requires the exposure of the whole body, strict regulations should be respected before any deployment. In fact, if the radiation concerns only one part of the human body, the radiation is considerably reduced. For instance, dental biometrics can be used for identification purposes. Using this modality, a routine exam requires only 0. 005 mSv. From each image,

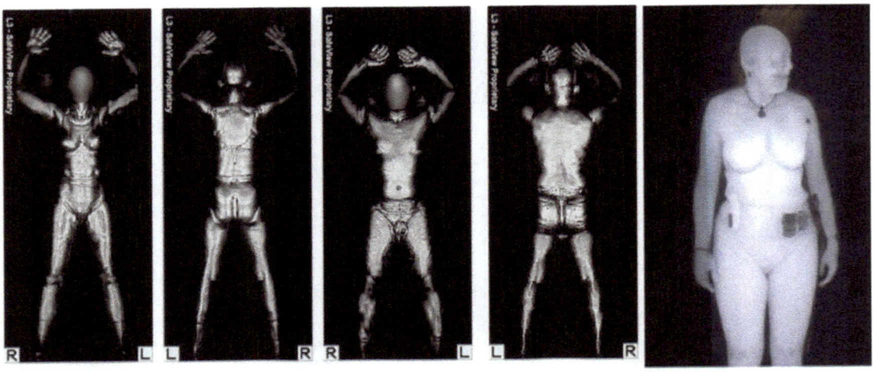

Fig. 4 Left/Imaging using a wave millimeter body scanner [2]. Right/backscatter X image [3]

features are extracted by considering geometry information (e.g. contours, shapes, etc.) (Fig. 5).

In the same context, hand X-ray imaging has successfully applied as a robust anti-spoofing human authentication system (Fig. 6) [8, 9]. In this work, safety aspects have been considered in order to define the optimal dose exposure. In fact, the National Council on Radiation Protection & Measurements (NCRP) recommendations, state that general-use systems should adhere to an effective dose of 0.1 μSv or less per scan, and can be used mostly regardless the number of individuals scanned or the number of scans per individual per year. Once the image acquired, the identification of an individual from the characteristics of the hand skeleton, requires the application of a few image processing operations to extract relevant features and therefore, characterize the hand phalanges, used as signature.

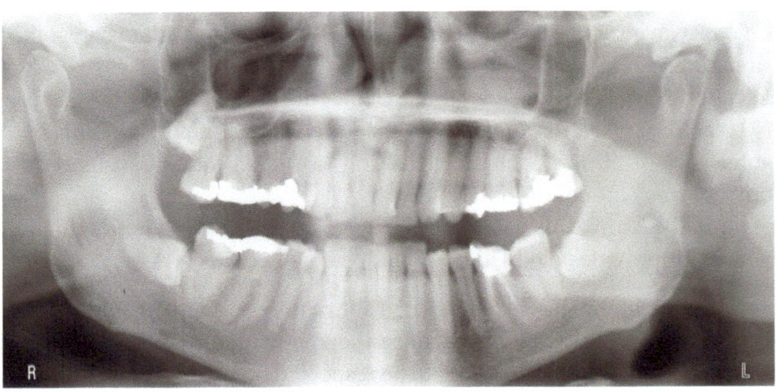

Fig. 5 Dental biometrics [4]. A radiation of about 0. 005 mSv is required

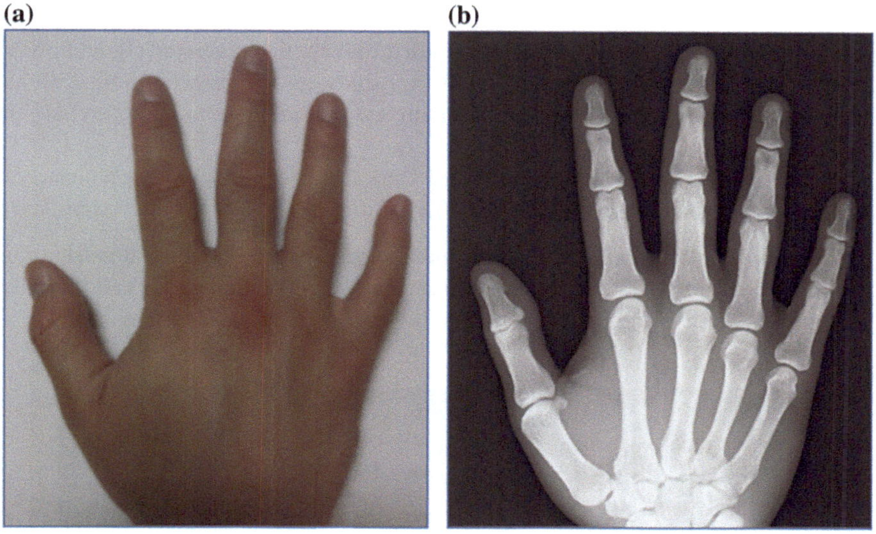

Fig. 6 X-ray hand imaging. On the left, an image acquired in the visible light. On the right, an image acquired through X-ray imaging

3.3 Magnetic Resonance Imaging (MRI) Approach

While X-ray imaging approach requires particular attention regarding safety aspects, Magnetic Resonance Imaging (MRI) does not use ionizing radiations which makes it potentially a good biometric modality candidate. Generally used in medical applications, MRI uses a strong magnetic field to visualize the inner human body. If this modality has been widely used in forensic applications, its use in biometrics for identification or verification is still under consideration. In fact, the main limitation of such modality is its cost and the required time to construct the inner image.

Many tentatives have been considered in MRI biometrics. for example, as an alternative to the conventional face recognition, MRI provides a full 3D image allowing not only the visualization of the face, but also the whole inner structures, including brain structures. Consequently, with this type of imaging, several 2D planes can be processed separately, namely: the sagittal section, the coronal section and the axial section. More specifically, if we focus only on the brain as a region of interest, studies showed that geometric shape of the cerebral cortex is unique for each individual. In fact, three types of cortical folding can be distinguished, such as: (1) the primary folding, characterized by a weak inter-individual variability, visible from the 16th week of gestation, (2) the secondary type of folding, characterized by an intermediate level variability which appears at about the 32nd week of gestation establishing the Gyrification degree of the cerebral cortex, and (3) the tertiary folding types, characterized by a high inter-individual variability, developed at about the 36th week of gestation.

When considering biometric applications, it has been shown in 2011 that folding patterns clearly describe individual variability, making the generated signature very interesting to identify individuals [10]. In this study, an accuracy up to 98.25% has been reached using Gabor wavelet for feature extraction of the brain codes [11, 10] (Fig. 7).

As an advanced approach, an extended work has been achieved for which a 3D processing schema has been used (Fig. 8), requiring the following main steps.

1. Curvilinear slices extraction: curvilinear slices are used from volumetric brain scans. Once projected as two-dimensional images, curvilinear slices allow planar representation of the cortex as well as the cortical folds as a texture image.

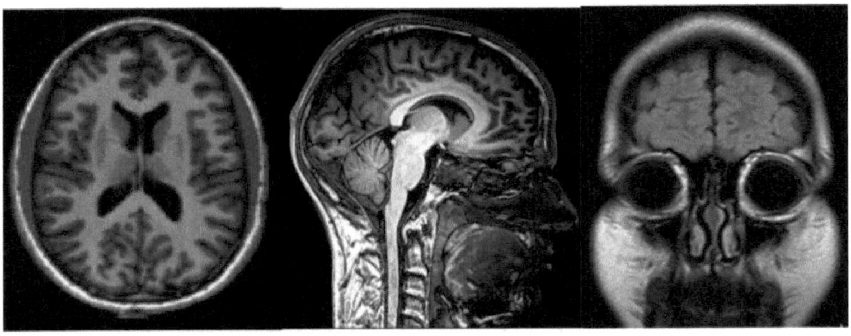

Fig. 7 Brain imaging showing Cerebral cortex surface 2D projection on coronal, axial and sagittal [6, 7] views

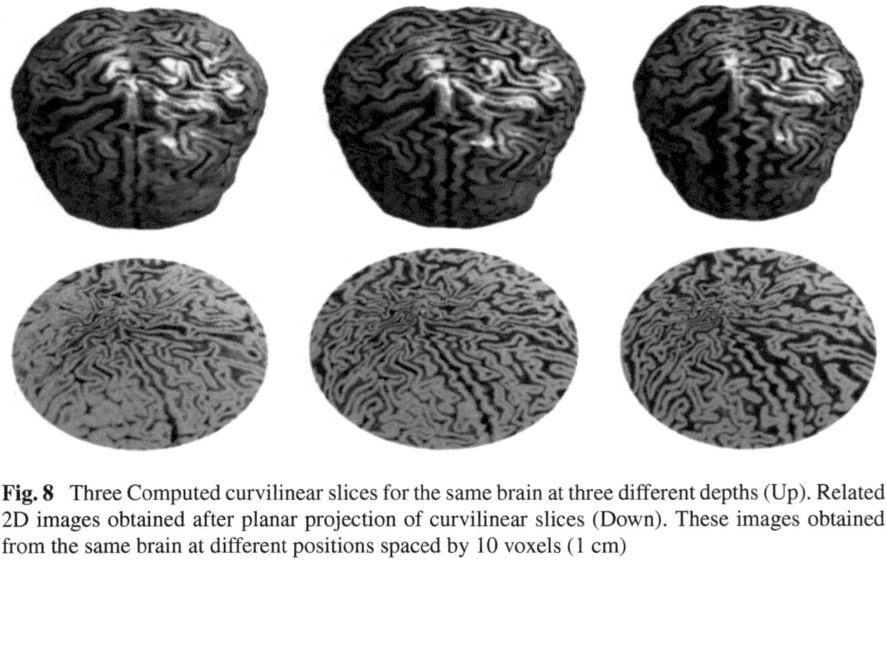

Fig. 8 Three Computed curvilinear slices for the same brain at three different depths (Up). Related 2D images obtained after planar projection of curvilinear slices (Down). These images obtained from the same brain at different positions spaced by 10 voxels (1 cm)

2. Gabor features extraction: this transform is used to extract textural features from projected curvilinear slices.
3. ACP: commonly used for dimensionality reduction by keeping only the most relevant characteristics of the Brainprints.
4. Performance evaluation: different classifiers have been used in order to highlight the uniqueness of human brain.

Through this research the representation of brain folds using curvilinear slices rather than standard planar slices produces symmetrical views of the cortex and allows low variability to be detected. In terms of performance, a promising correct classification rate of 99.64% has been reached. This rate can easily reaches 100% when using fusion techniques which consist of combining the results with those obtained by other modalities. For instance, brain features can be combined with the corresponding face features in the visible light or even using other multi-spectral imaging.

4 Conclusion

Providing a repeatable and efficient system to measure and extract features without interferences is a challenging task. Facing pose, illumination, expression and temporal variability as it is the case with face recognition systems, led to the consideration of hidden biometrics as a robust solution. For example, multispectral imaging such as IR and NIR are potentially interesting, but not as robust as X-ray that allows fast and relevant body inner imaging. However, X-ray imaging for biometrics should deeply be considered, and safety aspects should seriously be addressed through an optimal dose control. As an alternative, MRI is considered as a safe modality and provides high-resolution inner images. However, huge efforts should be made in order to make the imaging system more practical and suited for security applications. This may reduce automatically the cost of the acquisition process.

References

1. Taigman, Y., Yang, M., Ranzato, M., Wolf, L.: DeepFace: closing the gap to human-level performance in face verification. CVPR 2014
2. Transportation Security Administration (Public domain)
3. US Transportation Security Administration part of U.S. Department of Homeland Security (Public domain)
4. Timpo (Public domain)
5. Shoja, R. et al.: Infrared face recognition: a comprehensive review of methodologies and databases. Pattern Recognit. **47**(9), 2807–2824. ISSN: 0031-3203
6. Everyone's idle [CCBY-SA 2.0 (https://creativecommons.org/licenses/by-sa/2.0)]
7. © Nevit Dilmen [CC BY-SA 3.0 (https://creativecommons.org/licenses/by-sa/3.0)] Nevit Dilmen

8. Kabbara, Y., Shahin, A., Nait-Ali, A., Khalil, M.: An automatic algorithm for human identification using hand X-ray images. In: 2nd International Conference on Advances in Biomedical Engineering, pp. 167–170 (2013)
9. Kabbara, Y., Naït-Ali, A., Shahin, A., Khalil, M.: Hidden biometrie identification/authentication based on phalanx selection from hand X-ray images with safety considerations. In: International Conference on Image Processing Theory, Tools and applications (2015)
10. Aloui, K., Naït-Ali, A., Saber Naceur, M.: A novel approach based brain biometrics: some preliminary results for individual identification. In: 2011 IEEE Workshop on Computational Intelligence in Biometrics and Identity Management (CIBIM), Paris, pp. 91–95 (2011)
11. Aloui, K., Naït-Ali, A., Naceur, S.: A new useful biometrics tool based on 3D brain human geometrical characterizations. J. Signal Inf. Process. **3**(2), 198–207 (2012)

Retinal Image Processing in Biometrics

Rostom Kachouri, Mohamed Akil and Yaroub Elloumi

1 Introduction

Traditionally, the recognition of individuals is done either by the knowledge of passwords or answers to secret questions, either by an access badge or a Custom smart card. However, each of these methods suffers from a set of problems such as loss, forgetfulness and reproduction. Nowadays, the use of biometric recognition has become more and more a requirement for dealing with growing insecurity problems.

Generally, biometric characteristics are classified into three major modalities: behavioral, biological and morphological. The morphological modality may be based on specific physical traits such as fingerprint, face, palm print, retina, etc. The biological modality may be based on the analysis of biological traces such as odor, saliva or DNA. While the behavioral modality is based on the analysis of certain behaviors of people like the voice, the signature, the manner of walking, etc.

In this chapter, we are interested in one of the most morphological modalities, particularly, Retinal image processing in Biometrics. Indeed, Retinal is one of the safest modalities because of its protection inside the eye. Moreover, the retinal vascular network is a unique model of each individual and does not change throughout the life of the person. In addition, the retina offers a high level of recognition, which makes it suitable for high security applications thanks to its universality, its invariability over time and its difficulty to falsify.

R. Kachouri (✉) · M. Akil · Y. Elloumi
Gaspard Monge Computer Science Laboratory, ESIEE-Paris, University Paris-Est, Marne-la-Vallée, France
e-mail: rostom.kachouri@esiee.fr

Y. Elloumi
Medical Technology and Image Processing Laboratory, Faculty of Medicine, University of Monastir, Monastir, Tunisia

ISITCom Hammam-Sousse, University of Sousse, Sousse, Tunisia

A. Nait-ali (ed.), *Hidden Biometrics*, Series in BioEngineering,
https://doi.org/10.1007/978-981-13-0956-4_10

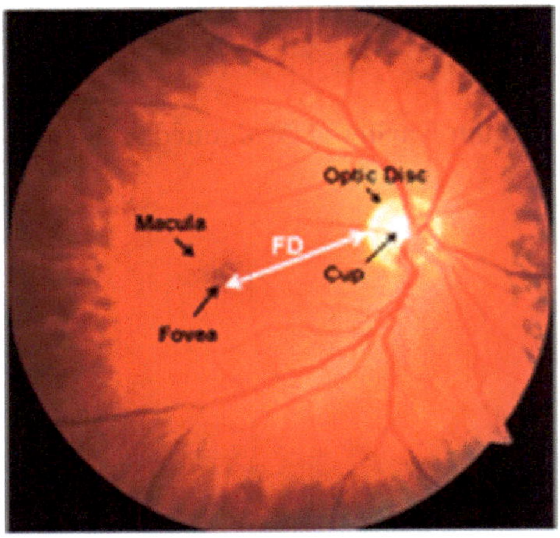

Fig. 1 Retina anatomy

The anatomy of Retina (Fig. 1) has a rich structure containing many characteristics and retinal components such as the optic disk, the cup, the macula and the fovea. In addition, the retina is traversed by blood vessels. The optic disc, where blood vessels converge, is characterized especially by the brightest area called Cup. On the other hand, the macula has the darkest part (fovea) of the retinal image. It is surrounded on each side by the vascular system from both upper and lower arteries. As illustrated in Fig. 1, the segment FD connects the fovea and the boundary of the optic disc. It is located on the line called the Raphe of the retina, which separates the inferior retinal region and the superior one. According to the retina anatomy (Fig. 1), the slice FD is equal to two Optic Disc diameters.

At first, Retinal Imaging plays a key role including the diagnosis and treatment of the ocular diseases [1]. It is in this area where much of the retinal capture devices developed both in field of research and in companies. Scientific and technological innovations in the design of retinal capture devices can contribute significantly to develop the Retina based identification systems.

The Retina based identification is perceived as the most secure method of authenticating an identity. Every eye has its own pattern of blood vessels. Thus, the retinal vasculature is a reliable biometric parameter, since each person's eye is known to have a unique arrangement of blood vessels in the retina [2, 3].

Moreover, retinal based identification provides a high level of accuracy. It is well suited for high-security applications. It is reputed to be very reliable and difficult to falsify. Figure 2 shows the different processing stages of the retina-based identification system.

The rest of this chapter is organized as follows: Sect. 2 presents an overview of existing Retinal imaging and capture devices. Retinal features for biometry and related works of existing retinal recognition systems are described in Sect. 3. In Sect. 4, we present some ocular pathologies affecting retinal components in biometric field. Finally, conclusion is drawn in the last section.

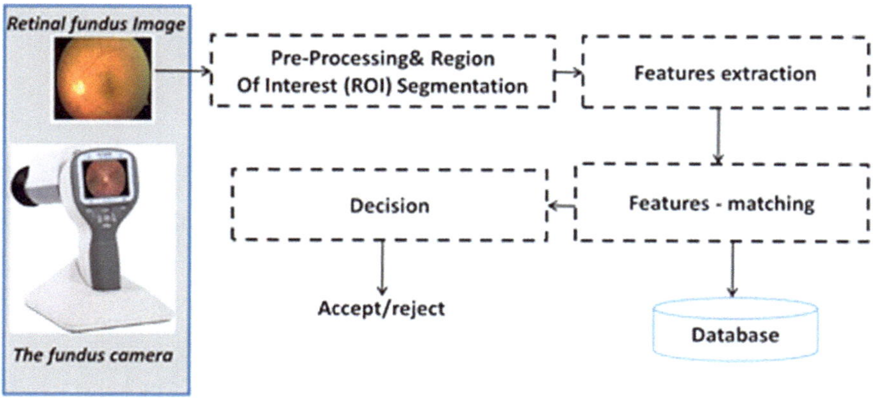

Fig. 2 Main steps of retina-based identification system

2 Retinal Capture Device

The retina is the sensory layer of the eye that allows vision. It is captured using a biometric scanner that uses a visible beam of light to trace a circular path on the retina to the back of the eye. Once the eyes are illuminated, a scan is performed to capture the features of the retina. So, in the beginning, this process is therefore invasive, uncomfortable, and difficult to implement. As a result, the acquisition systems have been improved over time. A near-infrared light source replaced the visible light source. The amount of energy radiated by this source as well as the acquisition time have decreased more and more with each new retinal capture device system. We present in this section, an overview of existing Retinal imaging and capture devices.

2.1 Retinal Imaging Modalities

Retinal imaging devices are primarily used in the earlier screening and diagnosis of ocular (retinal) diseases. The principal imaging technologies for the retina, are scanning laser ophthalmoscopy (Scanning laser ophthalmoscope—SLO) [4] and Optical Coherence Tomography (OCT) [5] and fundus camera imaging [6].

In the following, we present the main characteristics of these imaging modalities.

2.1.1 Scanning Laser Ophthalmoscope—SLO

The ophthalmoscope was introduced by Hermann Ludwig Ferdinand von Helmholtz in 1851. The Scanning Laser Ophthalmoscope (SLO) produces images at low light level using light of specific wavelengths. SLO creates a 2D image by scanning a laser

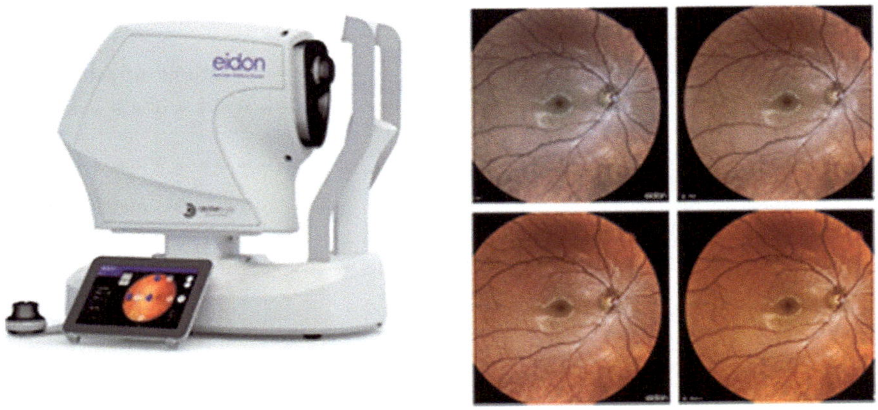

Fig. 3 EIDON's SLO system (https://www.centervue.com/) with 60° field in a single exposure and confocal view of the retina

beam across the 3D retinal surface instead of using a bright flash of white light, thus It acquires high contrast 2-D in-face retinal images with for example a resolution between 10 and 15 μm (Fig. 3).

Let us remember that the first Optical Coherence Tomography (OCT) system was designed at MIT Lincoln Laboratory in 1993.

2.1.2 Optical Coherence Tomography—OCT

OCT modality is a non-invasive imaging technique relying on low coherence interferometry to generate a cross-sectional volumetric image of ocular tissues. In 2001, ultrahigh resolution (UHR) OCT improves image quality and enables definitive visualization of individual retinal layers. Imaging can be performed with few μm (<7 μm resolution in [7]) and deep penetration (up to few millimeters). Some retinal diagnostic instrument can combine two (OCT and SLO—Confocal Scanning Laser Ophthalmoscope) or three functions (OCT, SLO and MP—Microperimetry, also called Fundus related perimetry). Microperimetry allows for exact topographic correlation between fundus details and its light sensitivity (Fig. 4).

2.1.3 Fundus Camera Imaging

The fundus camera Topcon TRC-50 V was conceived in 1983 by Tokyo Optical. Fundus camera is an optical system capable of illuminating and imaging the retina simultaneously. An image-capture device such as a digital camera is mounted on top of the optical system. The fundus camera provides a color photograph of retinal (Retinal Photography) surface with resolution between 7 and 20 μm (Fig. 5).

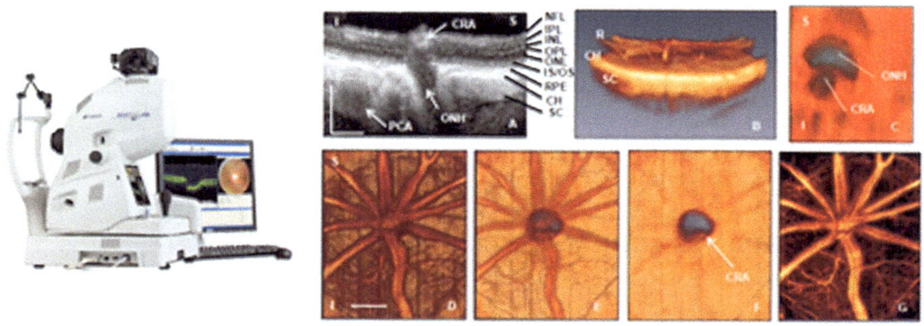

Fig. 4 Topcon 3D OCT-2000 (https://www.mediconsult.ch/) with digital non-mydriatic retinal camera—OCT structural image showing: layers and ONH anatomy, rat central retinal artery (CRA), choroidal microvasculature

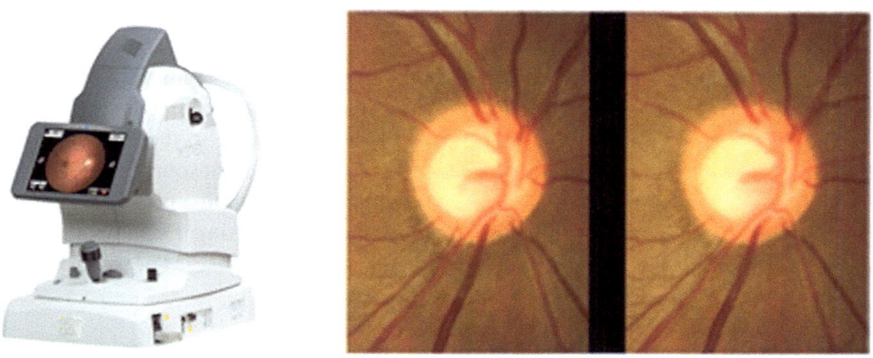

Fig. 5 Non-mydriatic auto fundus camera AFC-330 (http://www.nidek.fr/site/fr/)—image fundus of ONH and blood vessels

Fundus camera imaging generates a two-dimensional (2D) image is performed with a system that consists of a specialized low power microscope and an attached camera. After completing camera positioning the operator presses the shutter-release to fire a flash and creates the image. This image is an upright, magnified picture of the fundus with typical angles of view (30°, 45° or 60°) and with a desired magnification, depending on the system optics. Some modifications to these parameters are achievable through zoom or auxiliary lenses. For example, a 15° lens can provide 35 magnification, whereas a 140° wide angle lens captures a larger area of the fundus. A larger field of view (FOV) can be achieved by composing multiple images acquired at different fixation points. Most commonly, the retina is illuminated by white light and examined in full color. However, the imaging light can be filtered to remove red components, creating a red-free image with improved contrast of retinal and choroidal blood vessels and other structures.

The ever-increasing development of Fundus camera technology and its cost make the use of the non-mydriatic fundus cameras very appropriate in the field of retina identification. In this context, the choice of such device requires to us consider the main technical and inherent characteristics of each fundus camera such as field of view, View magnificence, View optical resolution, Minimum pupil diameter, Image resolution, etc.

2.2 Specification and Operation of the Retinal Capture Device

2.2.1 Working Principle of Fundus Camera

As proposed in [8], the fundus camera consists of an illumination system and an imaging system. The imaging system includes eyepiece, Beam splitter, analyzer, relay lens, objective lens and CCD. From annular source (i.e. light source) to the Beam splitter the path includes the linear polarizer and Condensing lens. The illumination system and imaging system share a common eyepiece and are connected by a beam splitter. The working principle is based on the emission of the beam from the annular source through the illumination system and the edge of pupil to the fundus tissue. Figure 6 shows a diagram block of the optical system with a light source (green LED) forming an annular light pattern, which was focused on the pupil and crystalline lens and defocused on the retina. The pupil imaging is sued for setting the subject's eye at the central position. The Retina Imaging part consists of two lenses. The first focusing

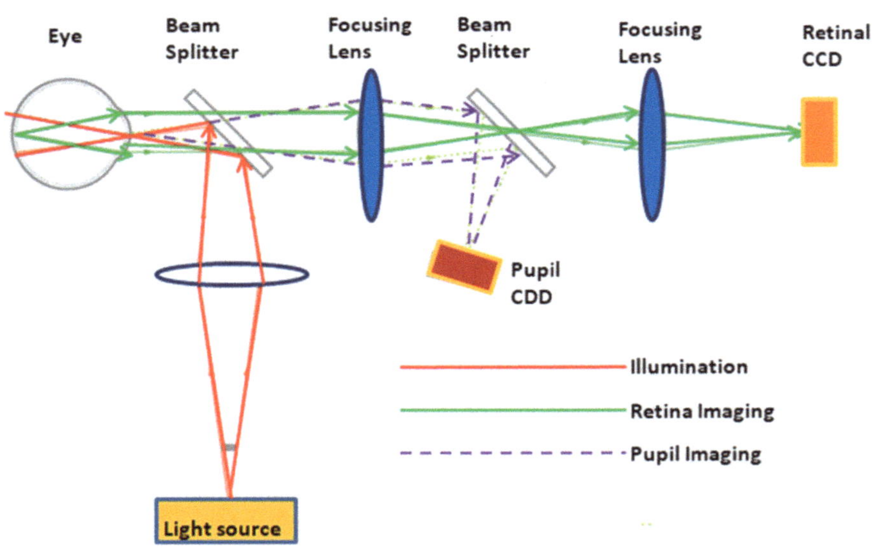

Fig. 6 Example of functional block diagram of fundus photography

lens provides a magnified secondary image in front of the second lens which in turn forms a further magnified image onto the retinal camera.

2.2.2 Main Specifications of Fundus Camera

Conventional fundus imaging, with a fundus camera typically utilizes a 30° or 35° field of view. This allows optimal visualization of the posterior pole, but the more peripheral retina is not captured. However, recent advances in optics have greatly extending the field of view, known as ultra-wide field (UWF) imaging. Non-mydriatic retinal camera provides ease of use, it is a non-contact, non-invasive technique for imaging the retina and optic disc using fundus camera. Non-mydriatic retinal cameras use infrared light to image the retina without requiring chemical dilation of the pupils (mydriasis). In addition, latest technologies of the Non-mydriatic retinal cameras are portable and easily transportable. Typically, a fundus camera will cover 45-60 degrees in one exposure. It is possible to combine multiple frames to achieve coverage of up to 140°, while ultra-wide field retinal capture imaging can cover up to 200° in a single exposure [9] (Fig. 7).

The **Field of View (FOV)** is the portion of the object that fills the camera's sensor. The **Working Distance (WD)** is the distance from the front of the lens to the object under inspection. The **Resolution** is the minimum feature size of the object that can be distinguished by the imaging system. The **Depth of Field (DOF)** is the maximum object depth that can be maintained entirely in acceptable focus. The **Sensor Size** is the size of a camera sensor's active area, typically specified in the horizontal dimension. The **PMAG** is the Primary Magnification of the lens is defined as the ratio between the sensor size and the FOV. [Imaging Resource Guide: Eo Edmund–optics–worldwide (https://www.edmundoptics.fr/resources/application-notes/imaging/understanding-focal-length-and-field-of-view/)] (Fig. 8).

Thus, it is important to establish an automatic mode for centering, autofocusing, capturing and monitoring the eye movement (from right eye to left eye). 3 fixation points such as Optic Nerf Head, central point and Macula and other fixation points to

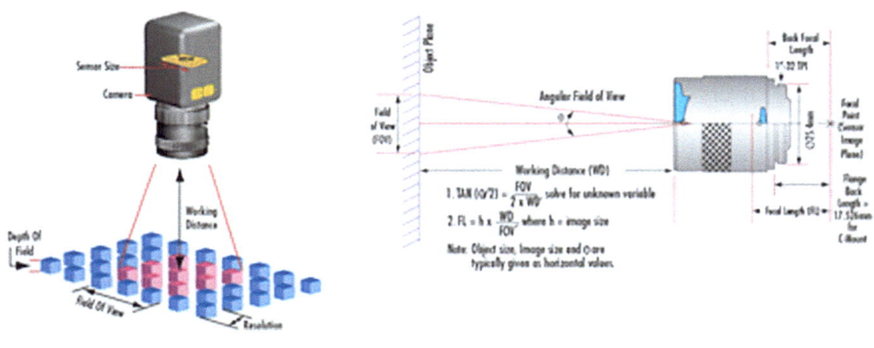

Fig. 7 Illustrations for some specifications of the camera device

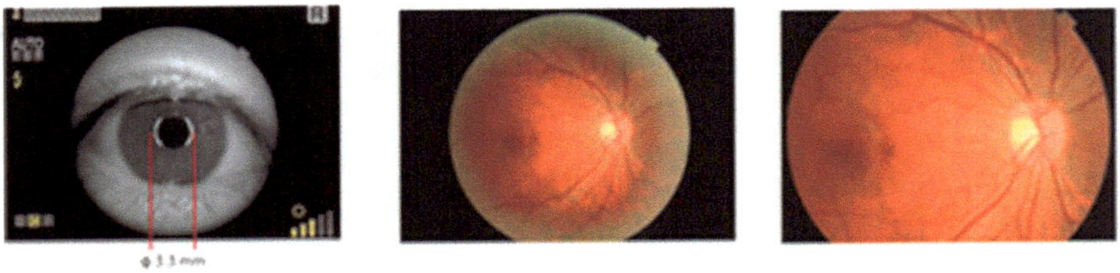

Fig. 8 Automatic detection of the pupil (with angle of 45°) and angle of 30° (with digital zoom)

take the peripheral images. For example, the TRC-NW400 has these characteristics and provides a "small pupil mode, with diameter of 3.3 mm.

2.2.3 Related Work and Fundus Camera Devices

The works related in [10–12] provide an exhaustive synthesis of digital fundus camera used in the context of ocular diagnosis [10, 11] and the technological advances [12].

This synthesis highlighted the advances of fundus photography for retinal screening as well as discusses the advantages, disadvantages, particularly by considering the impact of quality of image on the intended application.

The fundus camera devices described are: The iCam [fundus camera designed by Optovue, Inc. (Fremont, CA)], 3nethra (portable fundus camera designed and developed by Forus Inc), DIGITAL RETINOGRAPHY SYSTEM (Designed by CenterVue (Padova, Italy), digital retinography system (dRS), nonmydriatic fundus camera, requiring a pupil size of 4 mm), EasyScan (retinal imaging system developed by iOptics (Den Haag, The Netherlands), which works on the principle of scanning laser ophthalmoscopy), VISUCAM 200 (nonmydriatic fundus camera from Zeiss (Oberkochen,Germany)), NONMYD7 (nonmydriatic fundus camera from Kowa (Torrance, CA), magnification over 20° FoV to image the optic nerve head), CANON CR-2 (non-mydriatic digital retinal camera from Canon (Tokyo), low-power LEDs for illumination and photography, with autofocus, autofundus, and autocapture modes and a digital filter processor. Canon CR-2 provides 45° and 35° viewing angles in small pupil mode (pupil size, 3.3 mm). The California ultra-widefield retinal imaging device from Optos (Dunfermline, Scotland, United Kingdom) covers as wide as 200° or up to 82% of the retina capture in a single image.

2.2.4 Example of the Main Specifications of the ZEISS CLARUS 500's Fundus Camera

This fundus camera device captures true color imaging, high-resolution fundus autofluorescence (FAF) images–FAF-Blue and FAF-Green–and external eye images. The ultra-widefield mode captures a high-resolution image down to 7 μm and high-resolution details from the posterior pole to the periphery. All True Color images can be separated into red, green and blue channel images to help enhance the visual contrast of details in certain layers of the retina. The Field of View (measured from the center of the eye): Widefield (one image) 133°, Ultra-widefield (two images) 200°, Montage (up to six images) up to 267°. The other characteristics are: Resolution: Optical (7.3 μm), Minimum Pupil Diameter (2.5 mm), Working Distance (25 mm, it's patient's eye to front lens) and Light Sources: Red LED (585–640 nm), Green LED (500–585 nm), Blue LED (435–500 nm) and Infrared laser diode (785 nm).

2.2.5 Non-mydriatic Camera: Portable Device—Device-Based Smartphone

There as several type of non-mydriatic as the systems described in [13, 14]. Some of them are based-Smartphone camera devices, as for examples: PEEK Retina (Peek Vision, London), D-Eye (D-Eye; Pasadena, Calif. with undilated pupils, a field of view of 5–8° and the optic disc detection with pupils as small as 2 mm), HFC MicroClear is handled-portable with 45° field view, the ZEISS VISUSCOUT 100—Handheld Fundus Camera has a non-mydriatic operation, with color and red-free images and 40° field of view. The Bosch handy Fundus camera is Mydriatic and Non-Mydriatic camera with 40° wide field of view. Other Non-mydriatic camera devices are Canon CR2, Cobra CSO-272, Kowa 900, Nidek AFC330, and Topcon TRC-NW8. In 2013, the price of the fundus camera devices is 18,200 and 25,200 £ (Fig. 9).

In [12] the view and the main components of the optical system of the D-Eye module are described. "Retinal images are acquired using coaxial illumination and imaging paths thanks to a beam splitter (C). The blue arrow depicts the path of the light; red arrow depicts the path of fundus imaging. Device components are glass platelet (A) with imprinted negative lens (A), photoabsorbing wall (B), beam splitter

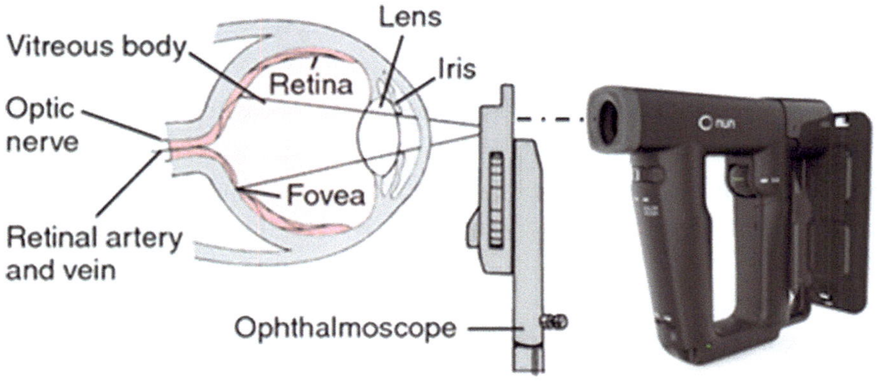

Fig. 9 Positioning the non-mydriatic camera device relative to the eye

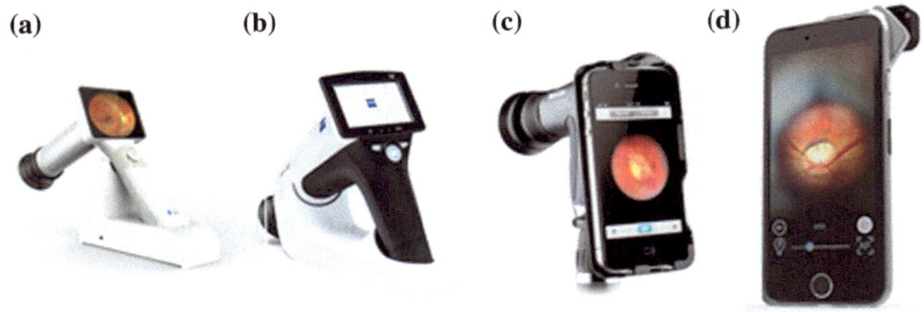

Fig. 10 Some non-mydriatic and portable fundus camera devices: **a** Horus DEC 200 hand held fundus camera (https://www.bibonline.co.uk/) with 40° manual focus portable fundus camera)—($3,995), **b** ZEISS VISUSCOUT 100—handheld fundus camera (https://www.zeiss.com/) ($ 9,950.00), **c** Volk iNview—iPhone fundus camera (https://www.foresightintl.com/) ($995), **d** the D-EYE retinal camera (https://www.d-eyecare.com/) ($435)

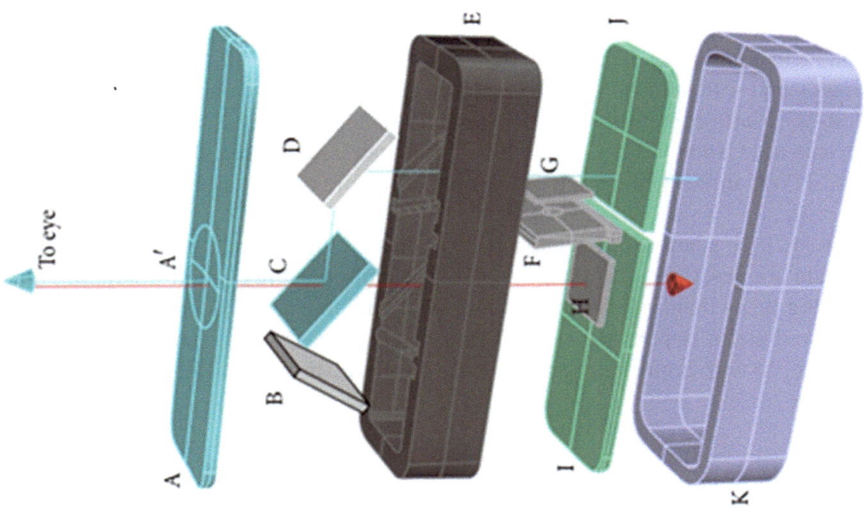

Fig. 11 The main components of the optical system of the D-Eye module [14]

(C), mirror (D), plastic case (E), diaphragm (F), polarized filters (G, H), flash and camera glass (J, I), and magnetic external ring (K)" [14] (Figs. 10 and 11).

3 Retinal Images in Biometrics

The retina is a safe biometric characteristic, where it is very difficult if not impossible to defeat. According to the evaluation criteria for biometric modalities, the Retina is Universal (present in all individuals); The arrangement of the Retinal vessels is

Unique from one individual to another, different between twins and stable during the life of individuals; Retina provides also a good Reliability and high Robustness against fraud, since it is an inner layer of the eye. In addition, it is also minimally exposed to injury, which minimizes errors of false acceptance or false rejection between individuals.

Biometric methods are always based on extracting and exploring retinal components in order to extract from them a feature set that allows person authentication. In this section, we are interested on which components and characteristics are aimed by biometric methods.

One approach has consisted in extracting blood vessels in the retinal fundus images and detecting characteristic points that serve as a biometric signature in order to authenticate a person's identity. In this approach, the person identification is based on the measure of similarity between blood vessel image generated from the Retinal fundus image as an input and reference images [15, 16].

Several research works rely on many different methods for automatic detection of the retinal vascular structure (vascular segmentation/Retinal vessel segmentation) in fundus camera images. Vascular segmentation allows the vessel detection, vascular branching points extraction and the measurement of parameters such as vessel diameter, bifurcation geometry and vascular tortuosity [17–20] (see Fig. 12).

In [21], the authors indicate that the number and the angles of major blood vessels in the area of the optic disc, their way to emerge from the optic nerve, the branching characteristics of the blood vessels and the size of the optic disc are unique. The proposed method leads to segment blood vessels in the region near to OD in order to generate unique identification code. Hence, it extracts the OD and then segment the blood vessels inside where result is modelled in binary image, as indicated in Fig. 13.

The work described in [19] aims to extract features related to the ridge endings and ridge bifurcations of the retinal vessel tree. In the first step, the level set extrinsic curvature (LSEC) is applied to the vessel tree in order to extract invariance properties which are ridges or valleys. After OD extraction, the method identified a set of points that are the ridge endings and bifurcations.

The work described in [22] informs that every bifurcation points and angles formed between each bifurcation point are different in every human being. This work leads

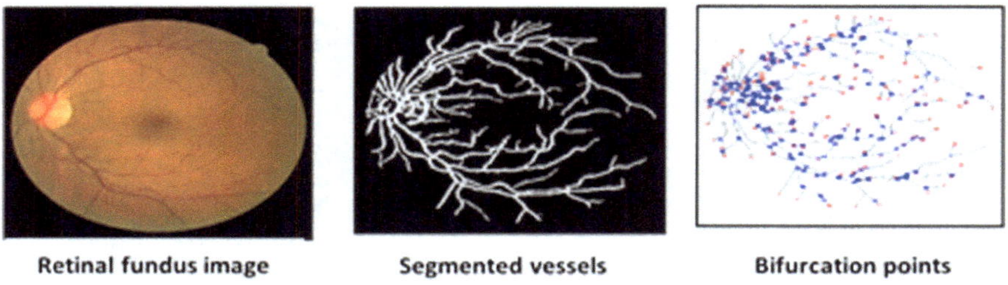

Fig. 12 Retinal fundus image, its segmented vessels and the bifurcation points

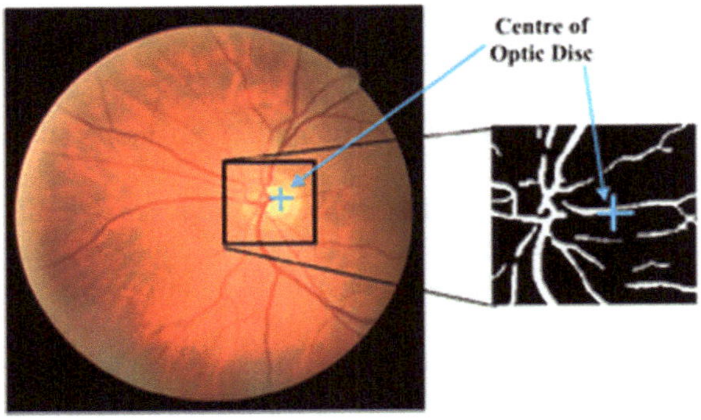

Fig. 13 Blood vessel segmentation in OD border

to apply an algorithm for bifurcation point detection, as illustrated in Fig. 14, in order to provide features related to the bifurcation angle.

In [16], the method leads to extract the OD in order to extract the blood vessel tree. The provided result allows identifying the person where the authentication consists at providing the same tree with an acceptable similarity. In [23], an angular partitioning mask is defined (Fig. 15) which leads to split the retina into angular sections using a defined degree. Thereafter, radial partitioning is performed by providing several concentric circles. Then, the vessel tree is segmented and then a morphological algorithm is applied for thinning vessels to preserve thicker and more significant vessels for identification and eliminating thinner ones. The authentication is performed by iteratively comparing vessel for each angular and radial sections.

In [24], the authors affirm that ending points and bifurcations are the key of retina recognition. However, there are certain vascular structures which contain invalid bifurcation and ending. The proposed method presents a new windowing technique for feature point validation in order to improve the accuracy of retinal vascular pattern recognition. The processing pipeline leads to segment vessel tree and applying

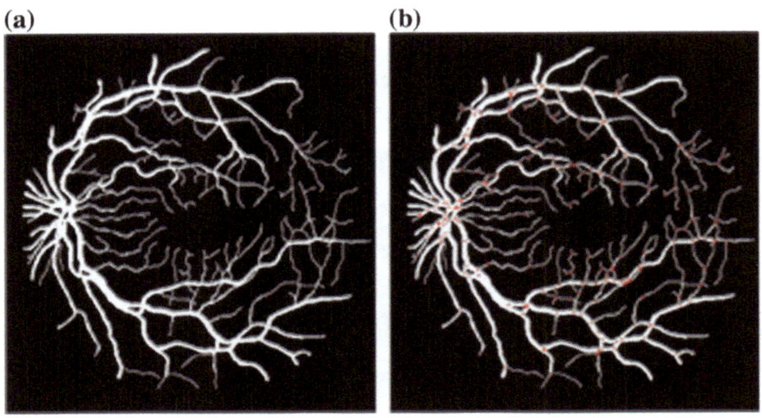

Fig. 14 **a** Segmented vessel tree; **b** vessel tree with bifurcation points

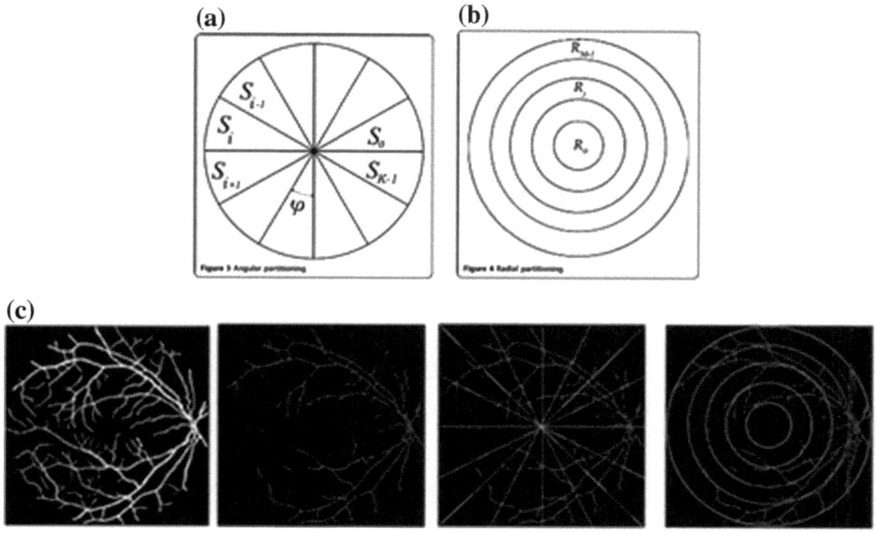

Fig. 15 **a** Angular partition; **b** radial partition; **c** processing pipeline of [16]

thinning. Thereafter, it eliminates spurs, breakages and short vessels in order to identifying ending point and bifurcation point.

Several supervised methods of blood vessels have been recently proposed. This kind of segmentation is generally based on the classification of pixels. Ricci and Perfetti [25] used the machine vector support classifier (SVM) to decide if a pixel belongs to the vascular network or not. They used two gray-level orthogonal line detectors from the target pixel to construct the characteristic vector. To improve the segmentation of the retinal blood vessels, we find methods based on neural networks that are more efficient and faster. Marin et al. [26] used a multi-layered neural network for the detection of retinal blood vessels, and Alonso et al. [27] used a faster type of neural networks: Deep Learning (CNN).

Figure 16 illustrates a synthesis of the results of some supervised (Fig. 16c) and unsupervised (Fig. 16d–f) retinal blood vessels segmentation methods. It shows a visual and qualitative comparison between manual expert segmentation (Fig. 16b) and the other automatic segmentations. Visually, we find that unsupervised methods based on the growth of regions as well as the adaptive filter allow extraction of the details of the finest blood vessels. These methods may result in the presence of a few redundant and irrelevant crossing points. While the Neural Network Based Segmentation (KNN) method extracts an incomplete vascular tree with some major retinal vessel discontinuities as shown in Fig. 16c.

We can see that most existing identification methods are based on the extraction of the Retinal blood vessels. However, these approaches may suffer from some disadvantages. First, information from the vascular system may be incomplete due to the discontinuity of some retinal segments in bifurcations and crosses. Otherwise,

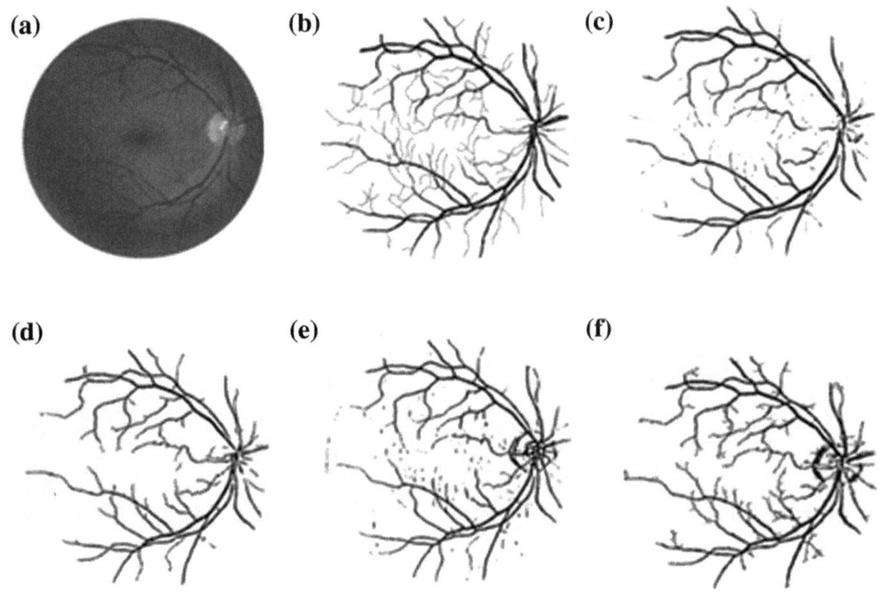

Fig. 16 Results of retinal blood vessels segmentation using different existing methods [28]: **a** the original retinal image, **b** the image segmented manually by experts, **c** using the supervised method KNN, **d** using mathematical morphology, **e** using the matched filter and **f** using a method based on the growth of regions

some diseases and ocular pathologies of the eye, can modify vascular characteristics or reveal microvascular, which cannot be extracted effectively.

4 Pathological Retinal Images in Biometrics

Biometric methods pretend always that the used retinal components are unique and hence allows person authentication. However, there are several ocular pathologies that may affect retinal components and features and hence distort biometry, such as cataract, glaucoma, diabetes, DMLA, hypertension and arteriosclerosis, etc. This may cause abnormalities and lesions like exudates and hemorrhages in the retina. In this section, we present several ocular pathologies that are characterized by a higher patient numbers, where disease grade retinal components used on biometry.

The glaucoma is a chronic ocular pathology that damage progressively the optic nerve head. It consists at significant rise of intraocular pressure. The patient number will achieve 76 million in 2020 and 111 million in 2040. It leads to damage nerve fibers and hence an excavation in the optic nerve head, such as modelled in Fig. 17a. Hence, the glaucoma leads to a progressive reduction of the neuro-retinal rim. Moreover, it occurs to clustering vessels in the border of OD. Those modifica-

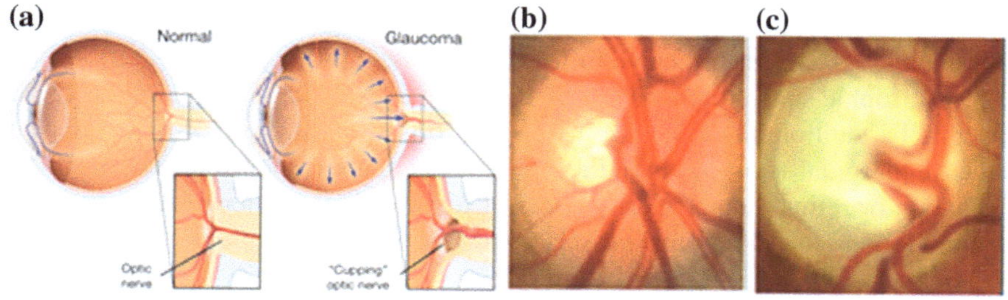

Fig. 17 **a** Glaucoma impact to the optic nerve head [29]; **b** healthy fundus image; **c** glaucoma fundus image

tions are deduced from the healthy and the glaucoma-diseased fundus images that are illustrated respectively in Fig. 17b, c.

Consequently, the glaucoma disease leads to shift blood vessels in the optic disc region. Moreover, bifurcations and their angles are adjusted with respect to the excavation. In addition, grouping vessels leads to superimpose them where may be modelled as unique vessel in the fundus image.

The diabetic retinopathy is a microvascular complication that may lead to vision lost. The World Health Organization indicates that 366 million persons can be affected by DR in 2030. Late stage in diabetic retinopathy is characterized by adding new abnormal vessels due to the lack of oxygen, as indicated in Fig. 18a.

Furthermore, Age-related macular degeneration (AMD) is an eye disease that primarily affects the elderly. The number of AMD patients could reach 196 million in 2020 and 288 million in 2040 [22]. It appears after the age of fifty and causing an alteration of macular retinal function and therefore central vision. Age-related Maculopathy (ALM) represents the early stage of the disease. It is characterized by the presence of alterations in the pigment epithelium of the macula or retinal complaints called exudates. At the back of the eye the exudates appear as whitish round lesions, of varying shape and size. The presence of an ALM carries a risk of developing AMD

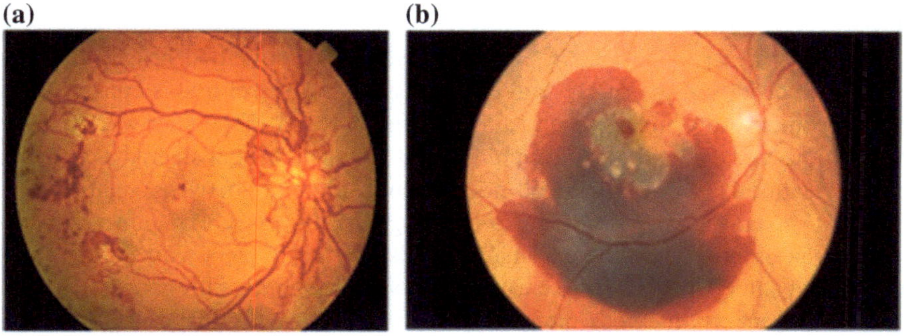

Fig. 18 **a** DR neovascularization; **b** neovascularization in wet AMD

at 5 years. There are two forms of age-related macular degeneration: an atrophic form (80% of cases), and an exudative form (20% of cases) [30].

The atrophic shape (cuttlefish shape) is in the form of large white lesions on the macula. In addition, the exudative form (wet form) is characterized by the appearance of choroidal neo-vessels in the macula, which are responsible for edema and exudation or retinal hemorrhages. The Wet form leads to a choroidal neovascularization. It consists at abnormal vessels that appears in the macula and then progress to neighbors' regions, as shown in Fig. 18b.

In both cases, those vessels emerge from main vessels and propagate in the whole retina. They are characterized by thin and torturous shapes, such as indicated in Fig. 18a, b which correspond respectively to the DR and the AMD. Moreover, neovascularization may bring to hemorrhage that are modeled as great red lesions in fundus images.

Accordingly, the neovascularization leads to modify vessel tree. Further, new vessels correspond to new bifurcation points where angles are diversified with respect to the tortuosity. Even biometry methods proceed to segment thick vessels, the higher density of added vessels avoid detecting thick ones correctly.

We find that generally all these retinal abnormalities presented attack the main elements of the retina namely the optic disk and the macula, which affects the description of the retina. As a result, they can interfere with differentiating between individuals and increasing false recognition rates in biometric field. One solution to overcome these problems of pathological retinal images may be to find the most stable region, even where retinal pathologies appears.

5 Conclusion

The retina is the sensory layer of the eye that allows vision. It is captured using a biometric scanner that uses a visible beam of light to trace a circular path on the retina to the back of the eye. The acquisition systems have been improved over time. We have presented in Sect. 2, an overview of existing Retinal imaging and capture devices.

Based on described works in Sect. 3, most biometry method based on retinal images proceed to segment OD and blood vessels. the vessel tree is considered as unique component which is used to authentication [16, 21, 23]. Other works extract bifurcations and ending points from blood vessels which are considered as personal unique features [19, 24]. To avoid noising and difference between segmentation methods, several methods proceed to thinning vessels before extracting biometry features, in order to preserve only main vessels [23, 24]. Others works aim to filter bifurcation and ending points to ensure a performant identification process [24].

These Biometric methods pretend always that the used retinal components are unique and hence allows person authentication. However, there are several ocular pathologies that may affect retinal components and features and hence distort biometry. In Sect. 4, we have presented several ocular pathologies that are characterized by a higher patient numbers, where disease grade retinal components used on biometry. The challenge here is to ensure efficient biometrics regardless of ocular pathology. In this context, Retinal Imaging plays also a key role including the diagnosis and treatment of the ocular diseases [1].

References

1. Abràmoff, M.D., Garvin, M.K., Sonka, M.: Retinal imaging and image analysis. IEEE Rev. Biomed. Eng. **1**(3), 169–208. https://doi.org/10.1109/rbme.2010.2084567 (2010)
2. Jain, A.K., Bolle, R., Pankanti, S.: Biometrics: personal identification in networked society. Springer Science & Business Media (1999)
3. Simon, C., Goldstein, I.: A new scientific method of identification. New York State J. Med **35**, 901–906 (1935)
4. Manivannan, A., Kirkpatrick, J.N.P., Sharp, P.F., Forrester, J.V.: Novel approach towards coulour imaging using scanning laser ophthalmoscope. Br. J. Ophthalmol. **82**(4), 342–345 (1998). https://doi.org/10.1135/bjo.82.4.342
5. Hermann, B., Fernandez, E.J., Unterhubner, A., Sattmann, H., Fercher, A.F., Drexler, W., Prieto, P.M., Artal, P.: Adaptative-optics ultrahigh-resolution optical tomography. Opt. Lett. **29**, 2142–2144 (2004)
6. DelHoog, E., Schwiegerling, J.: Fundus camera systems: a comparative analysis. Appl. Opt. **48**(2), 221–228 (2009)
7. Srinivasan, V.J., Huber, R., Gorczynska, I., Fujimoto, J.G.: High-speed, high-resolution optical coherence tomography retinal imaging with a frequency-swep laser at 850 nm. Opt. Lett. **32**(4) (2007)
8. Ma, C., Cheng, D., Xu, C., Wang, Y.: Design, simulation and experimental analysis of an anti-stray-light illumination system of fundus camera. In: Proceedings of SPIE—The International Society for Optical Engineering (2014)
9. Soliman, A.Z., Silva, P.S., Aiello, L.P., Sun, J.K.: Ultra-wide field retinal imaging in detection, classification, and management of diabetic retinopathy. Semin. Ophthalmol. **27**(5–6), 221–227 (2012)
10. Zhi, Zhongwei, Cepurna, William O., Johnson, Elaine C., Morrison, John C., Wang, Ruikang K.: Impact of intraocular pressure on changes of blood flow in the retina, choroid, and optic nerve head in rats investigated by optical microangiography. Biomed. Opt. Express **3**(9), 2220–2233 (2012)
11. Bernardes, R., Serranho, P., Lobo, C.: Digital ocular fundus imaging: a review. Ophthalmologica **226**, 161–181 (2011). https://doi.org/10.1159/000329597. Published online: 22 Sept 2011
12. Panwar, N., Lee, J., Chuan, T.S., Teoh, S., Huang, P., Keane, P.A., Richhariya, A., Lim, T.H., Agrawal, R.: Fundus photography in the 21st century—a review of recent technological advances and their implications for worldwide healthcare. Telemed. e-Health. https://doi.org/10.1089/tmj.2015.0068 (2015)
13. Russo, A., Delcassi, L., Morescalchi, F., Semeraro, F., Costagliola, C.: A novel device to exploit the smartphone camera for fundus photography. J. Ophthalmol (823139), 5 p. http://dx.doi.org/10.1155/2015/823139 (2015)
14. Russo, A., Morescalchi, F., Costagliola, C., Delcassi, L., Semeraro, F.: A Novel Device to Exploit the Smartphone Camera for Fundus Photography
15. Fraz, M.M., Remagnino, P., Hoppe, A., Uyyanonvara, B., Rudnicka, A.R., Owen, C.G., et al.: Blood vessel segmentation methodologies in retinal images: a survey. Comput. Methods Programs Biomed. **108**, 407–433 (2012)
16. Fukuta, K., Nakagawa, T., Hayashi, Y., Hatanaka, Y., Hara, Y., Fujita, H.: Personal identification based on blood vessels of retinal fundus images. Medical Imaging 2008, Image Processing,

Proceedings. of SPIE Vol. 6914, pp 1605–7422/08/$18 (2008). https://doi.org/10.1117/12.769330
17. Wang, L., Wong, T.Y., Sharrett, A.R., Klein, R., Folsom, A.R., Jerosch-Herold, M.: Relationship between retinal arteriolar narrowing and myocardial perfusion: multi-ethnic study of atherosclerosis. Hypertension **51**, 119–126 (2008). https://doi.org/10.1161/HYPERTENSIONAHA.107.09834
18. Dehghani, A., Ghassabi, Z., Moghddam, H.A., Moin, M.S.: Human recognition based on retinal images and using new similarity function. EURASIP J. Image Video Process. **2013**, 58 (2013)
19. Ortega, M., Marino, C., Penedo, M.G., Blanco, M., Gonzalez, F.: Biometric authentication using digital retinal images. In: Proceedings of the 5th WSEAS International Conference on Applied Computer Science, Hangzhou, China, April 16–18, pp. 422–427 (2006)
20. Modarresi, M., Oveisi, I.S., Janbozorgi, M.: Retinal identification using shearlets feature extraction. Austin Biometr. Biostat. **4**(1), id1035 (2017)
21. Singh, Anushikha, Dutta, Malay Kishore, Sharma, Dilip Kumar: Unique identification code for medical fundus images using blood vessel pattern for tele-ophthalmology applications. Comput. Methods Programs Biomed. **135**, 161–175 (2016)
22. Roy, N.D., Biswas, A.: Detection of bifurcation angles in a retinal fundus image. In: 2015 Eighth International Conference on Advances in Pattern Recognition (ICAPR) (2015)
23. Barkhoda, W., Akhlaqian, F., Amiri, M.D., Nouroozzadeh, M.S.: Retina identification based on the pattern of blood vessels using fuzzy logic. EURASIP J. Adv. Signal Process. 2011, 113 (2011)
24. Fatima, J., Syed, A.M., Usman Akram, M.: Feature point validation for improved retina recognition. In: 2013 IEEE Workshop on Biometric Measurements and Systems for Security and Medical Applications, Naples, Italy, 9–9 Sept 2013
25. Ricci, E., Perfetti, R.: Retinal blood vessel segmentation using line operators and support vector classification. IEEE Trans. Med. Imaging **26**, 1357–1365 (2007)
26. Marin, D., Aquino, A., Gegndez-Arias, M.E., Bravo, J.M.: A new supervised method for blood vessel segmentation in retinal images by using gray-level and moment invariants-based features. IEEE Trans. Med. Imaging **30**(1), 146–158 (2011)
27. Alonso-Montes, C., Vilarino, D.L., Penedo, M.G.: CNN-Based Automatic Retinal Vascular Tree Extraction. IEEE, pp. 61–64 (2010)
28. Zana, F., Klein, J.: Segmentation of vessel-like patterns using mathematical morphology and curvature evaluation. IEEE Trans. Image Process. **10**, 1010–1019 (2001)
29. Balakrishnan, U.: NDC-IVM: an automatic segmentation of optic disc and cup region from medical images for glaucoma detection. J. Innov. Opt. Health Sci. **10**(03), 1750007 (2017)
30. http://www.creteilophtalmo.fr/dmla/

From Motion to Emotion Prediction: A Hidden Biometrics Approach

Fawzi Rida, Liz Rincon Ardila, Luis Enrique Coronado, Amine Nait-ali and Gentiane Venture

Abstract In this chapter it will be discussed the capability of using motion recognition in order to predict the human emotion. Considered as a behavioral hidden biometrics approach, a specific system has been developed for this purpose wherein, several Machine-Learning approaches are considered, such as SVM, RF, MLP and KNN for classification and SVR, RFR, MLPR and KNNR for regression. The study highlights promising results in comparison to the state of the art.

1 Introduction

In early 1990s, the concept of Emotional Intelligence (EI) was introduced by the psychologists Jack Mayer and Peter Salovey. This concept is built on Social Intelligence which is the ability to understand one's own and other's internal state, motives, and behaviors to act toward them optimally on the basis of that information [1]. EI is related to a set of mental processes involving emotional information. These mental processes include: *Appraising and expressing emotions in the self and others, regulating emotions in the self and others and using emotions in adaptive ways.*

Appraising and expressing emotions are components of EI and a minimal competence of these skills is necessary for adequate social functioning allowing an appropriate socially adaptive behavior.

Emotional communication occurs through non-verbal channels like body movements and gestures, expressing emotions and recognizing emotions conveyed by these channels play a fundamental role in EI, as demonstrated by the experiments in [2]. Words and languages are not the only one of the communication channels that people use. Facial expressions, hand gestures, eye contact, voice tone and other

F. Rida
Soft Consulting, Paris, France

L. R. Ardila · L. E. Coronado · G. Venture (✉)
GV Lab, Tokyo University of Agriculture and Technology, Tokyo, Japan
e-mail: venture@cc.tuat.ac.jp

A. Nait-ali
Université Paris-Est, LISSI, UPEC, 94400 Vitry sur Seine, France

A. Nait-ali (ed.), *Hidden Biometrics*, Series in BioEngineering,
https://doi.org/10.1007/978-981-13-0956-4_11

cues can convey much more information than just words [3]. These elements are part of what it is known as nonverbal communication and are mostly uncontrolled and spontaneous. They can emphasize a verbal message, contradict it, call attention to important details and sometimes even distract from it. The ability to use and understand non-verbal cues seems to be a trait which is gradually developed from human birth, with very young children finding it difficult to read nonverbal cues that are easily read by older toddlers and children [4].

Biometrics rely on physical and behavioral characteristics. Lately researchers are more interested in hidden biometrics because of their characteristics of not being visible to outside world without scanning devices. Hidden biometrics are strong shields for spoofing attacks. Several works are made such as identify individuals from their MRI brain imaging [5] and from hand X-ray images and phalanx processing [6, 7]. In this chapter we introduce emotion as a hidden biometric related to behavioral characteristics, as claimed before about the importance of information driven by non-verbal communication, it is used to identify a person's motion and then predict the related emotion.

Emotion estimation and prediction has been a hot topic, where several machine learning technics have been used. We can distinguish three major emotion computing models: *discreet theory for categorical models, dimensional theory for dimensional models and appraisal theory for componential models* [8]. *Discrete models* describe an emotion state as discrete labels using the basic emotions as described by Ekman [9] some adjectives such as: "sad", "happy" etc. It is intuitive and simple but cannot express complex affective states. Introduced by Mehrabian and Russel [1], *dimensional models* consider an emotion state as a point in a continuous space. Therefore, dimensional theory can model subtle, complicated, and continuous affective behavior. Typically, an emotion state is covered by three dimensions: arousal (indicating the level of affective activation), valence (a measure of pleasure) and dominance (a measure of power or control). However, the three dimensions are enough for describing so many emotion states. At the same time, getting high quality dimensional labels is difficult. According to Plutchik theory [10], *Appraisal theory* attempts to detail the mental processes underlying the elicitation of emotions. It views an emotion state as a set of stimulus evaluation checks.

Emotion estimation and emotion prediction can be distinguished from (EI) as classification and regression problems, respectively. Usually, when it is a classification task the emotions are represented using the categorical model (using the basic emotions) several works have been made in [11–14] using different modalities such as face expression, speech signals, physiological signals and body movements.

The dimensional representation (values) is used to illustrate emotions while dealing with a regression, for instance in [15] they did a speech-based Continuous Emotion Prediction by Learning Perception Responses related to Salient Events which was a study based on Vocal Affect Bursts and Cross-Cultural Affect. In [16] Han et al. made a Prediction-based learning for continuous emotion recognition using speech signals. In [17] Baveye et al., made a comparison between deep learning and kernel methods to deal with the performance of emotion prediction in videos. In [18] Gunes and Piccardi made an automatic temporal segment detection with affect recognition from face and body display. In [19], Nicolaou et al. studied the spontaneous affect from multiple cues and modalities in valence-arousal space for a continuous prediction.

This chapter is organized as follow:

- A section for Emotion Prediction in which a large state of the art is made covering the most used modalities such as face expression, speech signals, physiological signals and body movements. Moreover, a brief definition and presentation of Self-assessment manikin and Pleasure, Arousal and Dominance models are made for better comprehension.
- The next section is about our proposed method for predicting Pleasure, Arousal and Dominance parameters from body movement in which we explain our approach and the followed procedure with the obtained results.
- The last section is a conclusion of our work and future perspective.

2 On Emotion Prediction

2.1 Facial Expression

Facial expressions are one of the most intuitive elements to analyze for Emotion Recognition (ER), especially in case of short distance from the subject. As a consequence, automatic facial gestures recognition has been studied in different researches, with excellent results.

In 1978, a work by Ekman and Friesen [20] established the Facial Action Coding System (FACS): a taxonomy, inspired by Hjortsjö [21], which describes facial movements by means of its elementary units, called Action Units (AUs). The first researches were based on static features. For example, [12, 13] have successfully trained neural networks with images to classify facial expressions. The use of dynamic features, such as facial movements, have overcome the limitations of methods based on static features, leading to better Correct Classification Rate (CCR) [14]. The necessity of performing the task in real-time has then focused the attention on video-based approaches. Dornaika and Davoine proposed an approach which allows the simultaneous retrieval of facial actions and expression using a particle filter adopting multi-class dynamics that are conditioned on the expression [22]. In [11], a head poses independent algorithm which relies on dynamic warping technique is presented. Despite the high CCR and robustness of different existing algorithms for facial expression recognition, this may not be the best approach for catching the actual emotion the analyzed subject is feeling. In fact, facial expressions can be controlled to some extent and no project addresses the problem of distinguishing between genuine and simulated facial expressions.

2.2 Speech Signals

Another branch of Emotion Recognition (ER) focuses on some parameters of the audio signals in verbal communication, such as tone of the voice and pitch: features related to these parameters are used as inputs to the emotion classification system. In [2], Luengo et al. has proposed approaches based on different features of prosody, spectral envelope and voice quality with features based on spectral envelope outperforming the prosodic ones. An attempt to create a speaker independent system has been proposed in [23] through an Iterative Feature Normalization (IFN) approach. Mao et al. have proposed a 2-learning-stage approach in which affect salient features for Speech Emotion Recognition (SER) are learned by using Convolution Neural Networks (CNNs) [24]. In [25], a multi-view supervised dictionary learning approach has been suggested where two techniques are proposed and application to SER using two feature sets are investigated. Correlation coefficient as the evaluation measure in predicting the continuous dimensional affects, the proposed methodology achieves the highest performance among the other methods. Recent researches demonstrated that multimodal approaches in which audio features are used along with facial features and body gestures overcomes the performances of mono-modal audio-based approaches [26].

2.3 Physiological Signals

The inner emotional state of a person can be identified also thanks to signals other than audio-visual ones. Recent researches address the problem of ER by analyzing physiological signals due to their direct or indirect relation with the emotional state of a person. Some example of physiological signals are Electromyogram (EMG) signal, Respiratory Volume (RV), Skin Temperature (SKT), Skin Conductance (SKC), Blood Volume Pulse (BVP) and Heart Rate (HR). Electroencephalogram (EEG) is one of the most studied physiological signals to effectively identify the emotions of a person [27]. Selection of physiological signals which are immediately linked to the internal state of a person like Fingertip blood oxygen saturation (OXY), Galvanic Skin Response (GSR), HR and breath gas are used by scientists in the field of Affecting Computing (AC) [28, 29]. The measurement of physiological signals, such as SKT and SKC, is also becoming extremely simple thanks to a new generation of portable sensors, like the Q sensor from Affectiva [30], allowing to collect data in real-life situations.

2.4 Body Movements

Since facial expressions and speech dominate during face-to-face interaction, these are the modalities that have been predominantly studied for automatic extraction of affect information, as discussed in the previous paragraphs. Nevertheless, considering body movement as a modality for AC is particularly suitable in situations where the affective state is estimated from a distance [31], to retrieve expressions which are less susceptible to social editing [32], and to communicate affective states which are easier conveyed through movement [33]. A possible emotion-aware categorization distinguishes:

Communicative movements show different emotions according to the specific gesture. They belong to daily life movements category that are meant to express an affective state [18, 31, 34, 35].

Functional movements are performed to accomplish a specific task, unrelated to the expression of affect [36, 37]. Many works have been done in the literature such as: Karg et al. [38] built automatic recognition models to examine the differences between inter-individual and person-dependent recognition accuracies for emotion categories using affective whole-body gait patterns. Same problem has been discussed by Bernhardt and Robinson [39] and by Gong et al. [21] where personal biases removed and remaining shown very interesting results where systems with personal biases are more accurate in correctly recognizing the involved emotion.

Moreover, Pollick et al. [40] have compared automatic affect recognition model performance with human recognition performance in distinguishing between angry and neutral state in knocking, lifting, and waving actions where a Multi-Layer Perceptron has been used to distinguish between anger and neutral status in the different motions. The results shown that MLP outperforms humans in affect recognition. Furthermore, Bernhardt and Robinson have studied the automatic recognition of happiness, anger, sadness and neutral status in non-stylized motions by using Support Vector Machines (SVMs) with polynomial kernel.

Artistic movements contain two categories known as: choreographed [41] and non-choreographed dancing [42]. This type of movements shows excess of expressions and can consist of a majority of movement types which that do not occur during daily life. Camurri et al. [43, 44] examined motion cues, such as Quantity of Motion (QoM) and Contraction Index (CI), in emotion expression in dance in four affective states: anger, fear, grief and joy.

Abstract movements for instance lifting the arms [45], this category differ from the previous categories because they are used neither to explicitly accomplish a task nor to communicate a meaning.

3 Predicting Pleasure, Arousal and Dominance Prediction from Body Gesture

3.1 Pleasure, Arousal and Dominance and Self-assessment Manikin Model

3.1.1 PAD Affective Model

Affective states can be represented using a set of distinct categorical labels, or a dimensional model. Categorical labels describe affective states based on their linguistic use in daily life. Different sets of categorical labels can be chosen depending on the study. Most frequently, happiness, sadness, fear and anger are included, a subset of the basic emotions [46]. A basic emotion is defined by a set of neural correlates in the brain, a specific set of bodily expressions, and a motivational component for action tendencies [46]. A popular set of basic emotions contains anger, happiness, sadness, surprise, disgust and fear [9, 47]. A common model applied in Affective Computing (AC) is the PAD-model, with the dimension's pleasure (or valence), arousal and dominance [45]. The PAD emotional state model is a psychological model developed by Albert Mehrabian and James A. Russell (1974 and after) to describe and measure emotional states. PAD uses three numerical dimensions to represent all emotions.

The PAD (Pleasure, Arousal, Dominance) model has been used to study nonverbal communication such as body language in psychology [20]. It has also been applied to consumer marketing and the construction of animated characters that express emotions in virtual worlds [45, 48]. Arousal corresponds to the level of activation, mental alertness, and physical activity. Dominance represents the amount of control over others and the surroundings versus feeling controlled by external circumstances. Categorical labels can be mapped to the continuous PAD space, for example, happiness, amusement and contentment are related to high pleasure, whereas anger, fear, and sadness are related to low pleasure [1, 49, 50]. A dimensional representation may relate more to the underlying physiological changes and Barrett suggests that categorical labels are a result of subjective categorization of emotions using conceptual knowledge, e.g. similar to color perception [51].

3.1.2 Self-assessment Manikin Model (SAM)

Picture-oriented instrument called the Self-Assessment Manikin (SAM) [52] was devised by Lang and Bradley to directly assess the pleasure, arousal and dominance associated in response to an object or event. SAM was originally implemented as an interactive computer program, and later was expanded to include a paper and pencil version for use in groups and mass screenings. Figure 1 depicts the paper and-pencil version of SAM illustrating its nonverbal, graphic depiction of various points along

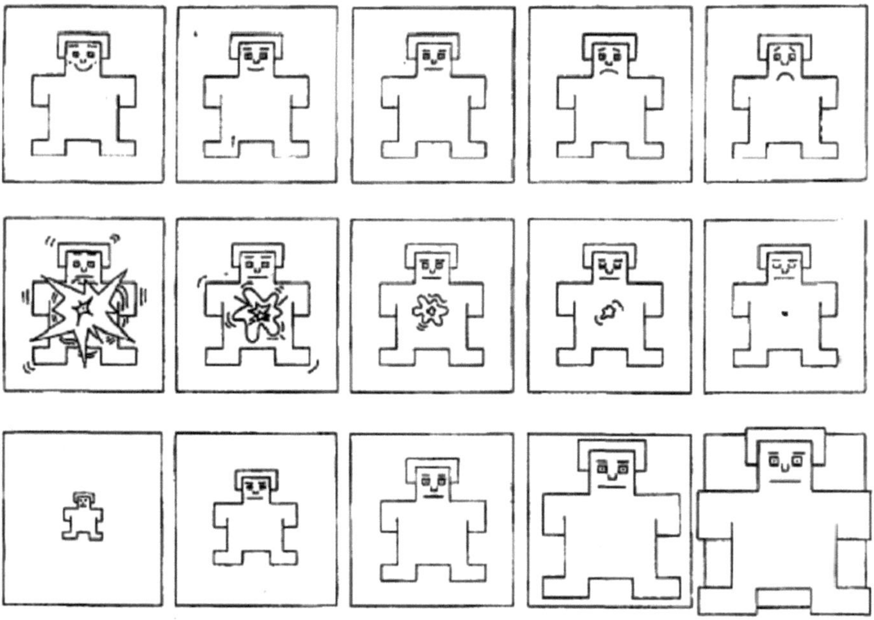

Fig. 1 The self-assessment manikin (SAM) used to rate the affective dimensions of valence (top panel), arousal (middle panel) and dominance (bottom panel). *Source* [52]

each of the three major affective dimensions. This figure is used in the experiments as well for measuring the emotions that the participants think robot may have. SAM ranges from a smiling, happy figure to a frowning, unhappy figure when representing the pleasure dimension, and ranges from an excited, wide-eyed figure to a relaxed, sleepy figure for the arousal dimension.

The dominance dimension represents changes in control with changes in the size of SAM. The SAM scale has been used to evaluate the degree of emotion conveyance. It is a visual questionnaire composed of three scales, each one corresponding to a dimension of the PAD space. It eliminates the problems related to verbal measures, and due to its visual nature, the participants can write their responses fast and intuitively.

3.2 *Motion Measurement Device*

In order to catch the motion data we used The Nintendo Wii which is a home video game console by Nintendo. It is characterized by a very intuitive interface that facilitate its use for a broad variety of people. One of the possible input devices of this console is the Wii Balance Board: a contact force measurement device which uses Bluetooth technology to communicate with the Nintendo Wii. The Wii Bal-

Fig. 2 Nintendo Wii balance board. *Source* [57]

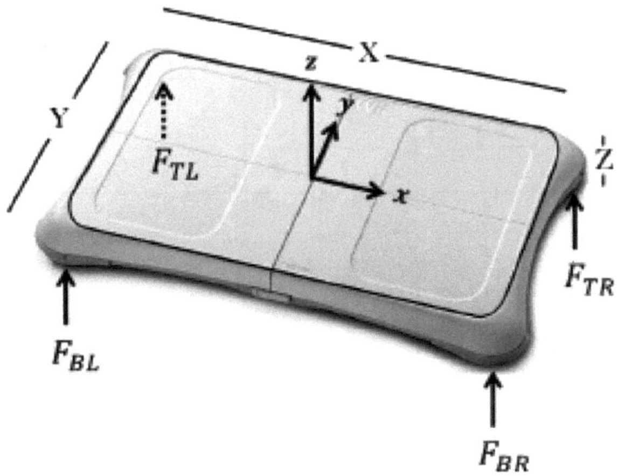

ance Board is a board controller which measures 511 mm long (X), 316 mm wide (Y) and 53.6 mm thick (Z). It can support up to 136 kg or 150 kg according to the considered version. It is equipped with 4 pressure sensors and the reference frame is described in Fig. 2. It has been used mainly for motion analysis in rehabilitation applications [53–55]. they've proven that. In [53], Clark et al. used the CoP position provided by the Wii Balance Board to study impaired standing balance. In [54], Shih et al. used data provided by Wii Balance Board to assess whether two persons with multiple disabilities would be able to control environmental stimulations using body swing (changing standing posture). In [56] they proposed an algorithm for motion recognition using the feature vector from force data obtained during a daily exercise program. They've concluded that there is a similarity of the clustering even for different candidates. The previous works have proven that the Wii Balance Board is a precise and reliable tool for assessing Center of Pressure (CoP) displacement, as well as the force along the vertical axis and the moment of force about the horizontal plane axes. Thus, this board can be used as an inexpensive and portable alternative to the classical force plates, since the performances are comparable.

3.3 Study Cases

In order to create a system that predicts emotions from movements we made a study about non-stylized motions. The human body has evolved not only to perform sophisticated tasks, but also to communicate affect and interpersonal attitudes. One possible distinction can be made between affect communicated through non-stylized and stylized motions. In a stylized motion the entirety of the movement encodes a particular emotion. Stylized motions normally originate from laboratory settings, where subjects are asked to freely enact an emotion without any constraints (even the motion itself is unspecified). They also arise from stylized dance. This research, however, concerns itself with the subtler aspects of non-stylized motions. How affect is communicated by the manner, in which every-day actions, such as knocking or walking, are performed is examined. This work focuses on the study of four different daily life motions: clapping, drinking, throwing and waving. These non-stylized motions fall into the category of functional movements because all of them have the primary goal of accomplishing some function or task.

The performances of the four motions listed above are studied when the subject is happy, sad, angry or neutral in order to be able to automatically recognize these four emotions from the execution of the motion.

Happiness, sadness and anger have been selected among the wide range of existing emotions since are considered to be a subset of the basic emotions 4 [46]; the neutral status, defined as the emotional status in which there is no predominant emotion, has been considered in order to offer an intermediate option in which no emotion is strong enough to affect the performance of the motion.

The experiments were conducted in an open space. A desktop computer was connected to a Nintendo Wii Balance Board (wireless Bluetooth connection). In order to acquire data for all the four motions (clapping, drinking, throwing and waving) performed in all the four emotional status (happy, sad, angry and neutral), acting school students has been enrolled as it shown in Fig. 3. Data related to 5 male actors between 23 and 26 years old has been collected. Table 1 shows the information about each single actor. The value of Handedness field in Table 1 indicates which hand (right or left) the subject prefers to use in a continuous scale 2 $[-1.0, +1.0]$ with 1.0 indicating «totally right handed» and -1.0 «indicating totally left handed». The given values are the result of the "Edinburgh inventory" [93]. Each actor has been asked to perform all the analyzed motion in all the selected emotional status. For each motion 60 samples were recorded, the total is 240 samples data were all samples are labeled as (motion, P, A, D).

Fig. 3 Representation of the throwing motion given through five frames taken from the video of one sample in the dataset. *Source* Original

Table 1 Gender, ago, handedness and nationality of the actors enrolled for the experiments

Subject	Gender	Age	Handedness	Nationality
Actor1	Male	23	+1.7	Japanese
Actor2	Male	24	+0.9	Japanese
Actor3	Male	23	+0.8	Japanese
Actor4	Male	26	+1.0	Japanese
Actor5	Male	23	+0.7	Japanese

3.4 Methodology and Experimental Results

In order to establish the hoped system, a multistage algorithm is made. The main parts of our system are the following: (i) Data acquisition, (ii) Data Processing, (iii) Segmentation, (iv) Motion Classification and finally (v) Emotion prediction.

3.4.1 Data Acquisition

The Nintendo Wii Balance Board sensor is used to acquire the data. This sensor can provide the displacement along the x and y axes of the CoP and the vertical force along the z axis.

3.4.2 Data Processing

Feature scaling is used in order to bring all the values between the range [0, 1]. The normalized input $\hat{x}$ is then computed as

$$\hat{x} = \frac{x - x_{min}}{x_{max} - x_{min}}$$

with x the input vector, x_{max} and x_{min} constants representing the minimum and the maximum of the input vector respectively.

Interpolation and resampling are needed to deal with the difference in the sampling frequency.

Feature extraction: using PCA the number of principal components has been selected in such a way that the new coordinates keep 99% of the variance of the original data.

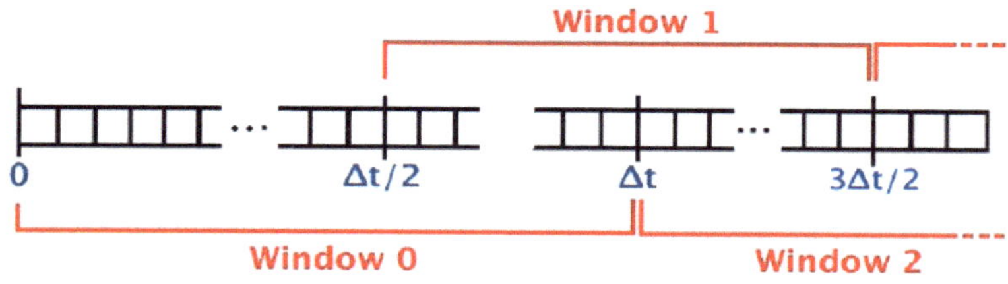

Fig. 4 Intuitive representation of data window's creation; each square represents the data related to one-time step. All the data arrived in the interval [t, t + Δt] are gathered in the same window. *Source* [41]

3.4.3 Segmentation

The data are segmented such that the portion related to the specific motion is isolated from the rest. Using this technic, we distinguish between data related to motion and data related to idle time. So instead of analyzing each time step separately, it is convenient to use overlapped moving window to regroup all the data within a certain time interval Δt [58] as it show in Fig. 4.

3.4.4 Motion Classification

In order to make the recognition of motions, popular classifiers are used such as: SVM, RF, KNN and MLP. Moreover, Uniform grid-search and CV have been used to explore the hyper-parameters space.

Motion Classification Results

The obtained results for motion classification are shown in Table 2.

All classifiers' performances go above 0.89 of accuracy (89% of CCR). The best performance among the off-line algorithms is obtained by MLP with a score of (0.97) and a standard deviation of (0.0504). By analyzing the misclassification cases, each classifier performs great on different motion.

3.4.5 Emotion Classification

Same as in classification we tried several affective regressors offered by scikit-learn [59] such as: SVR, RFR, KNNR and MLPR to predict jointly the values of three out-

Table 2 Summary of the results obtained for motion classification task with the different classifiers

Algorithms	KNN	SVM	RF	MLP
Accuracy	0.89	0.96	0.96	0.97
Standard deviation	0.06	0.05	0.05	0.05

puts (Pleasure, Arousal and Dominance) and exactly same procedure in classification was followed.

Emotion Prediction Results

The results obtained for emotion prediction, where we compared all the previous algorithms are shown in Table 3. Furthermore, we made a comparison between the best regressors and human's predictions. The used metric to evaluate the performance of our regressors is the explained variance. We can clearly see from the results in Table 3, among the considered regressors, the easiest motion to analyses remains "throwing", while the most difficult ones seem to be "drinking" and "waving". The difficulty of " clapping " seems to be medium and not relevant enough. The KNN regressor outperform the prediction rate of the other methods in "clapping" with 0.7281 explained variance score. MLP regressor outperform the others in "drinking" with 0.5183 explained variance score. Random Forest Regressor outperform the rest of the methods in "Throwing" with a score of 0.8574 of explained variance score. For "Waving" the best score is obtained with MLP regressor with 0.6225 explained variance score.

Human Baseline

In order to be able to interpret in the best way the results obtained in this project for emotion prediction task, a baseline is determined by asking a group of people to recognize the emotion from the videos of the motions. 20 video samples of motions have been processed by blurring the face of the actors both to preserve their privacy and for avoiding people to recognize the emotions from facial expressions. 10 people have been asked to fill a questionnaire made in Power-point Microsoft office (see Fig. 5) which contains the blurred video samples along with a self-assessment manikin pictures model with PAD values from $[-1, 1]$.

After filling the values of PAD, they were asked some questions:

- Their knowledge about SAM and PAD models?
- What they think about their PAD estimations?

The results of these experiments are shown in Tables 4, 5, 6 and 7. The used metrics in this task are the Mean Absolute Error (MAE) and the Mean Squared Error (MSE). A comparison of Human Prediction ss AI methods is made. To calculate the errors rate between humans, we made 2 groups of 5 persons (5 vs. 5) and we compared their predictions using the previous metrics (MAE, MSE). In "clapping"

Table 3 Summary of emotion prediction, using the explained variance score

Motion	RF	KNN	MLP	SVR
Clapping	0.6358	0.7281	0.7158	0.6038
Drinking	0.5164	11.1	0.5183	0.4604
Throwing	0.8574	0.6631	11.5505	0.7136
Waving	0.6042	0.5625	0.6225	0.5447

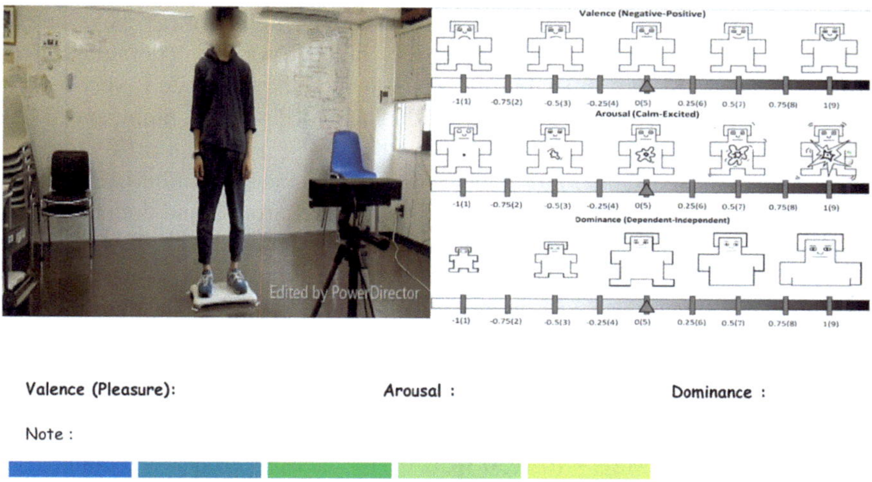

Fig. 5 A sample from the questionnaire. *Source* Original

Table 4 Summary of emotion prediction results KNN Regressor versus Human in Clapping motion, the used metrics are MAE and MSE

PAD	P_{KNN}	A_{KNN}	D_{KNN}	P_{HUMAN}	A_{HUMAN}	D_{HUMAN}
Mean absolute error	0.15	0.15	0.23	0.30	0.32	0.34
Mean squared error	0.03	0.04	0.07	0.16	0.17	0.19

Table 5 Summary of emotion prediction results MLP Regressor versus Human in Waving motion, the used metrics are MAE and MSE

PAD	P_{MLP}	A_{MLP}	D_{MLP}	P_{HUMAN}	A_{HUMAN}	D_{HUMAN}
Mean absolute	0.29	0.16	0.25	0.38	0.30	0.38
Error mean squared error	0.12	0.04	0.13	0.26	0.15	0.23

Table 6 Summary of emotion prediction results RF Regressor versus Ifnman in Throwing motion, the used metrics are MAE and MSE

PAD	P_{RF}	A_{RF}	D_{RF}	P_{HUMAN}	A_{HUMAN}	D_{HUMAN}
Mean absolute	0.16	0.10	0.10	0.37	0.34	0.33
Error mean squared error	0.05	0.016	0.017	0.22	0.193	0.18

Table 7 Summary of emotion prediction results RF Regressor versus Human in Drinking motion, the used metrics are MAE and MSE

PAD	P_{MLP}	A_{MLP}	D_{MLP}	P_{HUMAN}	A_{HUMAN}	D_{HUMAN}
Mean absolute error	0.32	0.21	0.28	0.30	0.32	0.34
Mean squared error	0.13	0.07	0.09	0.16	0.17	0.19

(see Table 4) for humans, pleasure and arousal scored a high error rate MAE ≈ 0.31 for both and MSE ≈ 0.16 and 0.17 respectively, higher error rate in dominance with MAE ≈ 0.34 and MSE ≈ 0.19. For the KNN method the error rate is very low for the 3 outputs for pleasure, arousal MAE ≈ 0.15 and MSE ≈ 0.03, dominance MAE ≈ 0.23 and MSE ≈ 0.07.

For "waving" (see Table 5) for humans the error rate is high with MAE $\approx$ (0.38, 0.30, 0.38) and MSE $\approx$ (0.26, 0.15, 0.23) for pleasure, arousal and dominance. for Multi layers perceptrons MLP method the error rate is lower with MAE $\approx$ (0.29, 0.16, 0.25) and MSE $\approx$ (0.12, 0.04, 0.13) for pleasure, arousal and dominance. Also, for "throwing" (see Table 6) for humans the error rate is high with MAE $\approx$ (0.37, 0.34, 0.33) and MSE $\approx$ (0.22, 0.193, 0.18) for pleasure, arousal and dominance. For RF method the error rate is low with MAE $\approx$ (0.16, 0.10, 0.10) and MSE $\approx$ (0.05, 0.016, 0.017) for pleasure, arousal and dominance.

Finally, in "drinking" (see Table 7) for humans the error rate is high with MAE $\approx$ (0.30, 0.32, 0.34) and MSE $\approx$ (0.16, 0.16, 0.19) for pleasure, arousal and dominance. for MLP method the error rate is higher in pleasure with MAE ≈ 0.32 and MSE ≈ 0.18 and for arousal and dominance it is lower MAE ≈ 0.21 and 0.28 and MSE ≈ 0.07 and 0.09.

According to these results, the average humans MAE is ≈ 0.3 while the average in Artificial intelligence AI methods prediction MAE is ≈ 0.2, at the same time, the average humans MSE is ≈ 0.19 while the average AI methods prediction MSE is ≈ 0.1. These results reveal and show that the proposed algorithm outperform humans in the task of emotion prediction by MAE ≈ 0.1 and MSE ≈ 0.09 for the selected study cases and confirms the relevance of the obtained results.

4 (N − 1). Conclusion

In this work we developed a system with the capability to predict the emotional status of humans in four daily life motions (non-stylized motions) which are clapping, drinking, throwing and waving, by analyzing the body movement.

As claimed before about the importance of body gestures as non-verbal communication channel for carrying information the internal state of the subject, this topic has not been exhaustively explored in the literature especially with the use of continuous model for emotion representation. This fact is the main motivation of this work in which a novel approach of emotion prediction through body gesture analysis has been presented.

Three main stages can be identified as the main steps that bring to the final emotion prediction: motion segmentation, motion classification and emotion prediction. To accomplish the classification and regression tasks four popular Machine Learning techniques, SVM, RF, KNN and MLP, have been used. The results of the motion classification task show good results (above 0.9 of accuracy), while the results of emotion prediction for clapping, throwing and waving show medium and poor results and good results for drinking (0.86 explained variance).

Even if the proposed methods outperform the human perception, these results still not satisfying in the emotion prediction task which show the hard task of emotion prediction and the need of a higher level of abstraction algorithms such as deep learning methods for instance RNNs that deal with the temporal data.

The use of a deep neural network like RNN, allows a higher level of abstraction, it is crucial for accomplishing the complex task of emotion prediction which is shown in our new proposed paper in which we present a complete sophisticated system based on a robot that react to human motions, the environments and its internal states.

References

1. Russell, J.A., Mehrabian, A.: Evidence for a three-factor theory of emotions. J. Res. Pers. **11**(3), 273–294 (1977)
2. Luengo, I., Navas, E., Hernáez, I.: Feature analysis and evaluation for automatic emotion identification in speech. IEEE Trans. Multimed. **12**(6), 490–501 (2010)
3. Baron, R.A., Branscombe, N.R., Mynhardt, J.C.: Social Psychology. Pearson (2014)
4. Liebal, K., Carpenter, M., Tomasello, M.: Young children's understanding of markedness in non-verbal communication. J. Child Lang. **38**(4), 888–903 (2011)
5. Aloui K., Nait-Ali, A., Saber, N.M.: 2011 IEEE Workshop on Computational Intelligence in Biometrics and Identity Management (CIBIM), pp. 91–95
6. Kabbara, Y., Shahin, A., Nait-Ali, A., Khalil, M.: An automatic algorithm for human identification using hand X-ray images. In: 2013 2nd International Conference on Advances in Biomedical Engineering (ICABME), pp. 167–170 (2013)
7. Kabbara, Y., Nait-Ali, A., Shahin, A., Khalil, M.: Hidden Biometric Identification/Authentication based on Phalanx Selection from Hand XRay Images with Safety considerations. In: The fifth International Conference on Image Processing Theory, Tools and Applications, Orleans (2015)
8. Noroozi, F., et al.: Survey on emotional body gesture recognition. arXiv preprint arXiv:1801.07481 (2018)
9. Ekman, P.: Psychol. Rev. **99**(3), 550–553 (1992)
10. Plutchik, R.: The nature of emotions: human emotions have deep evolutionary roots, a fact that may explain their complexity and provide tools for clinical practice. Am. Sci. **89**(4), 344–350 (2001)
11. Dornaika, F., Moujahid, A., Raducanu, B.: Facial expression recognition using tracked facial actions: classifier performance analysis. Eng. Appl. Artif. Intell. **26**(1), 467–477 (2013)
12. Raheja, J.L., Kumar, U.: Human Facial Expression Detection from Detected in Captured Image Using Back Propagation Neural Network (2010)
13. Su, M.-C., Hsieh, Y., Huang, D.-Y.: A simple approach to facial expression recognition. In: Proceeding WSEAS 2007 (2007)
14. Zhang, L., Tjondronegoro, D.: Facial expression recognition using facial movement features. IEEE Trans. Affect. Comput. **2**(4), 219–229 (2011)
15. Gamage, K.W., Dang, T., Sethu, V., Epps, J.: Speech-Based Continuous Emotion Prediction by Learning Perception Responses Related to Salient Events: A Study Based on Vocal Affect Bursts and Cross-Cultural Affect
16. Han, J., Zhang, Z., Ringeval, F., Schuller, B.: Prediction-Based Learning for Continuous Emotion Recognition in Speech
17. Baveye, Y., Dellandréa, E., Chamaret, C., Chen, L.: Deep Learning vs. Kernel Methods: Performance for Emotion Prediction in Videos
18. Gunes, H., Piccardi, M.: Automatic temporal segment detection and affect recognition from face and body display. IEEE Trans. Syst. Man Cybern. Part B (Cybern.) **39**(1), 64–84 (2009)

19. Nicolaou, M.A., Gunes, H., Pantic, M.: Continuous prediction of spontaneous affect from multiple cues and modalities in valence-arousal space. IEEE Trans. Affect. Comput. **2**(2), 92–105 (2011)
20. Friesen, E., Ekman, P.: Facial action coding system: a technique for the measurement of facial movement. In: Palo Alto (1978)
21. Hjortsjö, C.-H.: Man's face and mimic language. Studen litteratur (1969)
22. Dornaika, F., Davoine, F.: Simultaneous facial action tracking and expression recognition in the presence of head motion. Int. J. Comput. Vis. **76**(3), 257–281 (2008)
23. Busso, C., et al.: Iterative feature normalization scheme for automatic emotion detection from speech. IEEE Trans. Affect. Comput. **4**(4), 386–397 (2013)
24. Mao, Q., et al.: Learning salient features for speech emotion recognition using convolutional neural networks. IEEE Trans. Multimed. **16**(8), 2203–2213 (2014)
25. Gangeh, M.J., et al.: Multiview supervised dictionary learning in speech emotion recognition. IEEE/ACM Trans. Audio Speech Lang. Process. (TASLP) **22**(6), 1056–1068
26. Lu, K., Jia, Y.: Audio-visual emotion recognition using boltzmann zippers. In: 2012 19th IEEE International Conference on Image Processing (ICIP), pp. 2589–2592. IEEE (2012)
27. Rozgíc, V, Vitaladevuni, S.N., Prasad, R.: Robust EEG emotion classification using segment level decision fusion. In: 2013 IEEE International Conference on Acoustics, Speech and Signal Processing (ICASSP), pp. 1286–1290. IEEE (2013)
28. Lakens, D.: Using a smartphone to measure heart rate changes during relived happiness and anger. IEEE Trans. Affect. Comput. **4**(2), 238–241 (2013)
29. Vinola, C., Vimaladevi, K.: A survey on human emotion recognition approaches, databases and applications. ELCVIA Electr. Lett. Comput. Vis. Image Anal. **14**(2), 24–44 (2015)
30. MIT Technology Review. Sensor detects emotions through the skin. https://www.technologyreview.com/s/421316/sensor-detects-emotions-throughthe-skin/. Last accessed on Web. 8 Aug. 2018
31. Kapur, A., et al.: Gesture-based affective computing on motion capture data. In: International Conference on Affective Computing and Intelligent Interaction, pp. 1–7. Springer (2005)
32. Ekman, P., Friesen, W.V.: Detecting deception from the body or face. J. Pers. Soc. Psychol. **29**(3), 288 (1974)
33. Aviezer, H., Trope, Y., Todorov, A.: Body cues, not facial expressions, discriminate between intense positive and negative emotions. Science **338**(6111), 1225–1229 (2012)
34. Ravindra De Silva, P., et al.: Towards recognizing emotion with affective dimensions through body gestures. In: 7th International Conference on Automatic Face and Gesture Recognition, 2006. FGR 2006, pp. 269–274. IEEE (2006)
35. Shan, C., Gong, S., McOwan, P.W.: Beyond facial expressions: learning human emotion from body gestures. In: BMVC, pp. 1–10 (2007)
36. Ma, Y., Paterson, H.M., Pollick, F.E.: A motion capture library for the study of identity, gender, and emotion perception from biological motion. Behav. Res. Methods **38**(1), 134–141 (2006)
37. Venture, G., et al.: Recognizing emotions conveyed by human gait. Int. J. Soc. Robot. **6**(4), 621–632 (2014)
38. Karg, M., Kuhnlenz, K., Buss, M.: Recognition of affect based on gait patterns. IEEE Trans. Syst. Man Cybern. Part B (Cybern.) **40**(4), 1050–1061 (2010)
39. Bernhardt, D., Robinson, P.: Detecting affect from non-stylised body motions. In: International Conference on Affective Computing and Intelligent Interaction, pp. 59–70. Springer (2007)
40. Pollick, F.E., et al.: Estimating the efficiency of recognizing gender and affect from biological motion. Vis. Res. **42**(20), 2345–2355 (2002)
41. Camurri, A., Mazzarino, B., Volpe, G.: Expressive interfaces. Cognit. Technol. Work **6**(1), 15–22 (2004)
42. Park, H., et al.: Emotion recognition from dance image sequences using contour approximation. In: Joint IAPR InternationalWorkshops on Statistical Techniques in Pattern Recognition (SPR) and Structural and Syntactic Pattern Recognition (SSPR), pp. 547–555. Springer (2004)
43. Camurri, A., Lagerlöf, I., Volpe, G.: Recognizing emotion from dance movement: comparison of spectator recognition and automated techniques. Int. J. Hum Comput Stud. **59**(1–2), 213–225 (2003)

44. Camurri, A., et al.: Multimodal analysis of expressive gesture in music and dance performances. In: International Gesture Workshop, pp. 20–39. Springer (2003)
45. Castellano, G., Villalba, S.D., Camurri, A.: Recognising human emotions from body movement and gesture dynamics. In: International Conference on Affective Computing and Intelligent Interaction, pp. 71–82. Springer, Berlin (2007)
46. Tracy,J.L., Randles, D.: Four models of basic emotions: a review of Ekman and Cordaro, Izard, Levenson, and Panksepp and Watt. Emot. Rev. **3**(4), 397–405 (2011)
47. Ekman, P., Friesen, W.V.: A new pan-cultural facial expression of emotion. Motiv. Emot. **10**(2), 159–168 (1986)
48. Ratneshwar, S., Mick, D.G., Huffman, C.: Introduction: the "why" of consumption. In: The Why of Consumption, pp. 21–28. Routledge (2003)
49. Havlena, W.J., Holbrook, M.B.: The varieties of consumption experience: comparing two typologies of emotion in consumer behavior. J. Consum. Res. **13**(3), 394–404 (1986)
50. Mikels, J.A., et al.: Emotional category data on images from the international affective picture system. Behav. Res. Methods **37**(4), 626–630 (2005)
51. Barrett, L.F.: Solving the emotion paradox: categorization and the experience of emotion. Pers. Soc. Psychol. Rev. **10**(1), 20–46 (2006)
52. Bradley, M.M., Lang, P.J.: Measuring emotion: the self-assessment manikin and the semantic differential. J. Behav Ther. Exp. Psychiatr. **25**(1), 49–59 (1994)
53. Clark, R.A., et al.: Validity and reliability of the Nintendo Wii Balance Board for assessment of standing balance. Gait Posture **31**(3), 307–310 (2010)
54. Shih, C.-H., Shih, C.-T., Chu, C.-L.: Assisting people with multiple disabilities actively correct abnormal standing posture with a Nintendo Wii balance board through controlling environmental stimulation. Res. Dev. Disabil. **31**(4), 936–942 (2010)
55. Venture, G., Yabuki, T., Kinase, Y., Berthoz, A., Abe, N.: Using Dynamics to Recognize Human Motion (2016)
56. Yabuki, T., Venture, G.: Motion classification and recognition using only contact force. In: Proceedings of the IEEE/RSJ International Conference on Intelligent Robots and Systems, pp. 4251–4256. Hamburg, Germany (2015)
57. Web Site: Wii Balance Board. http://www.mdpi.com/1424-8220/14/10/18244. Last accessed on Sun. 12 Aug 2018
58. Babcock, B., Datar, M., Motwani, R.: Sampling from a moving window over streaming data. In: Proceedings of the Thirteenth Annual ACMSIAM Symposium on Discrete Algorithms. Society for Industrial and Applied Mathematics, 17 Aug 2018, pp. 633–634 (2002)
59. Scikit-learn: Scikit-Learn—Machine Learning in Python. http://scikit-learn.org/. Last accessed on Sat. 18 Aug 2018
60. Gong, L., et al.: Recognizing affect from non-stylized body motion using shape of Gaussian descriptors. In: Proceedings of the 2010 ACM Symposium on Applied Computing, pp. 1203–1206. ACM (2010)
61. Izui, T., et al.: Expressing emotions using gait of humanoid robot. In: 2015 24th IEEE International Symposium on Robot and Human Interactive Communication (RO-MAN), pp. 241–245. IEEE (2015)
62. Karg, M., et al.: Body movements for affective expression: a survey of automatic recognition and generation. IEEE Trans. Affect. Comput. **4**(4), 341–359 (2013)
63. Mayer, J.D., DiPaolo, M., Salovey, P.: Perceiving affective content in ambiguous visual stimuli: A component of emotional intelligence. J. Pers. Assess. **54**(3–4), 772–781 (1990)
64. Mehrabian, A.: Nonverbal Communication. Routledge, 2017
65. Mehrabian, A.: Pleasure-arousal-dominance: a general framework for describing and measuring individual differences in temperament. Curr. Psychol. **14**(4), 261–292 (1996)
66. Ng, A.: Online Lecture Notes-Machine Learning. Stanford University
67. Oldfield, R.C.: The assessment and analysis of handedness: the Edinburgh inventory. Neuropsychologia **9**(1), 97–113 (1971)
68. Thorndike, R.L., Stein, S.: An evaluation of the attempts to measure social intelligence. Psychol. Bull. **34**(5), p. 275 (1937)

69. Walk, R.D., Walters, K.L.: Perception of the Smile and other Emotions of the Body and Face at Different Distances (1988)
70. ZeroMQ: Zeromq—distributed messaging. http://zeromq.org/. Last accessed

MIX
Papier aus verantwortungsvollen Quellen
Paper from responsible sources
FSC® C105338

If you have any concerns about our products,
you can contact us on
ProductSafety@springernature.com

In case Publisher is established outside the EU,
the EU authorized representative is:
Springer Nature Customer Service Center GmbH
Europaplatz 3, 69115 Heidelberg, Germany

Printed by Libri Plureos GmbH
in Hamburg, Germany